PRAISE FOR *LEADERSHIP MOMENTS FROM NASA*

"NASA teams motivated by extraordinary leaders have succeeded in overcoming unprecedented challenges to accomplish their mission. *Leadership Moments* captures the essence of these successful NASA leaders and tells their stories in an exciting, insightful and inspiring way."

— JOE ROTHENBERG, FORMER NASA ASSOCIATE
ADMINISTRATOR FOR SPACE FLIGHT

"Leaders in any business can learn from Dave Williams, a great and successful leader in a dangerous endeavor where incorrect decisions or ineffective inspiration can lead to death. It's not systems that launch people into space; it's great leaders and the people they inspire, who use systems wisely, that launch people into space. Learn from Dave Williams and lead your company to greatness!"

— CAPTAIN JIM WETHERBEE, U.S. NAVY (RET.),
VETERAN SPACE SHUTTLE COMMANDER

"An extraordinary collection of leadership insights. For anyone who wishes to be inspired and instructed by leaders who have balanced extreme risk with historical reward, this book is for you!"

— SCOTT HALDANE, RETIRED PRESIDENT AND CEO OF
YMCA CANADA AND RIDEAU HALL FOUNDATION

"If you enjoy stories about bold vision and achieving great results in challenging environments, I highly recommend *Leadership Moments from NASA*. The authors do a masterful job of connecting Project Apollo's dots with unique insights into the leaders and teamwork that created one of the most significant accomplishments of all time."

— IAN GRAHAM, FOUNDER AND TEAM MENTOR
AT STARTUP ACCELERATOR THE CODE FACTORY

"*Leadership Moments from NASA* is a deep dive into NASA's pioneering work in space exploration and leadership. The leadership strategies outlined apply to not only technology endeavors, but any situation. Whatever new frontier you are boldly headed into, this book is a resource for accomplishing the impossible."

— LOREN PADELFORD, GENERAL MANAGER OF
REVENUE AND VICE PRESIDENT OF SHOPIFY PLUS

LEADERSHIP MOMENTS FROM NASA

ACHIEVING THE IMPOSSIBLE

DR. DAVE WILLIAMS

*Astronaut and Former Director of Space
& Life Sciences at NASA*

AND ELIZABETH HOWELL, PhD

Published by ECW Press
665 Gerrard Street East
Toronto, Ontario, Canada M4M 1Y2
416-694-3348 / info@ecwpress.com

Cover design: David Drummond

LIBRARY AND ARCHIVES CANADA CATALOGUING IN
PUBLICATION

Title: Leadership moments from NASA : achieving the
impossible / Dr. Dave Williams (astronaut and former
director of Space & Life Sciences at NASA), and
Elizabeth Howell, PhD.

Names: Williams, Dave (Dafydd Rhys), 1954– author. |
Howell, Elizabeth, 1983– author.

Identifiers: Canadiana (print) 20210118784 | Canadiana
(ebook) 20210119098

ISBN 978-1-77041-604-8 (Hardcover)
ISBN 978-1-77305-718-7 (PDF)
ISBN 978-1-77305-717-0 (ePUB)
ISBN 978-1-77305-719-4 (Kindle)

Subjects: LCSH: United States. National Aeronautics
and Space Administration. | LCSH: Leadership—United
States—Case studies. | LCGFT: Case studies.

Classification: LCC HD57.7 .W55 2021 | DDC
658.4/092—dc23

This book is funded in part by the Government of Canada. *Ce livre est financé en partie par le gouvernement du Canada.*
We also acknowledge the support of the Government of Ontario through the Ontario Book Publishing Tax Credit, and
through Ontario Creates.

 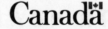

PRINTED AND BOUND IN CANADA

PRINTING: FRIESENS 5 4 3 2 1

MIX
Paper from
responsible sources
FSC® C016245
www.fsc.org

To all the leaders of the
space station partner agencies
for your dedication to making
humans a spacefaring species.

In memory of George M. Low,
an extraordinary leader.
— DAVID R. WILLIAMS

To the leaders and teams of the Apollo program,
who first inspired my interest in space.
— ELIZABETH HOWELL

TABLE OF CONTENTS

INTRODUCTION

"Houston, Tranquility Base, the Eagle has landed." July 20, 1969 — a day that will stand forever in history. With the advent of television, more people were watching the NASA lunar landing than any other event in history. It had been eight years since President John F. Kennedy proclaimed that NASA would send humans to the Moon and return them safely to Earth before the end of the decade. "Not because it is easy, but because it is hard." With that proclamation began one of the most incredible stories of leadership, teamwork and risk management in history. It takes courage and a relentless commitment to excellence to achieve the impossible. Even with today's space exploration capabilities, many wonder how NASA was able to accomplish this seemingly impossible feat, successfully achieving Kennedy's goal within the decade. It wasn't easy.

Many who dreamed of exploring space believed that it would be impossible. The televised Mercury, Gemini and Apollo missions and the many articles in *LIFE* and *National Geographic* magazines captured the imaginations of the young and old. It was clear that there were risks associated with space exploration; to push the boundaries of the unknown was not something to be taken lightly. In January 1967, the crew of Apollo 1 perished. Not in space but on the launchpad, in a fire lasting 90 seconds. The crew trapped inside the Apollo capsule had no chance for survival and NASA suffered the first loss of a spaceflight crew. When asked about risk in December 1966, Commander Gus Grissom responded, "You sort of have to put that out of your mind. There's always a possibility that you can have a catastrophic failure, of course; this can happen on any flight; it can happen on the last one as well as the first one. So, you just plan as best you can to take care of all these eventualities, and you get a well-trained crew and you go fly." A month later he, Roger Chaffee and Ed White would perish in the tragic fire.

Space exploration is the story of people working together through triumph and tragedy. Gene Kranz, now famous as the lead flight director during Apollo 13, responded to the Apollo 1 fire by calling a meeting of his staff in mission control three days after the accident. Not mincing words, he said, "We were too 'gung-ho' about the schedule and we blocked out all of the problems we saw each day in our work. Every element of the program was in trouble and so were we." With a steely gaze and short crew cut, Kranz, an aerospace engineer and former fighter pilot, embodied the NASA culture. "From this day forward, Flight Control will be known by two words: tough and competent," he said, looking each team member in the eye. "Tough means we are forever accountable for what we do or what we fail to do. We will never again compromise our responsibilities. . . . Competent means we will never take anything for granted . . . mission control will be perfect. When you leave this meeting today you will go to your office and the first thing you will do there is to write *Tough and Competent* on your blackboards. It will never be erased. Each day when you enter the room, these words will remind you of the price paid by Grissom,

White and Chaffee. These words are the price of admission to the ranks of mission control." The team left the meeting and refocused their efforts on one of the most significant achievements in history: sending humans to the Moon.

Galvanized in front of their televisions, 650 million viewers worldwide watched as the crew of Apollo 11 undocked the lunar module (LM) from the command service module (CSM) as Neil Armstrong and Buzz Aldrin began their descent to the lunar surface. Quiet and unassuming, Neil Armstrong had already developed a reputation for his humility and his performance under stress. His recovery of an out-of-control spacecraft caused by a stuck thruster on Gemini VIII and his understated reaction to ejecting from the "flying bedstead" when it started to plummet out of control a few hundred feet in the air had already given him legendary status in the astronaut office. Now, the next three hours would determine if he and Buzz Aldrin would become the first humans to land on the surface of the Moon.

"You are go for undocking," the CAPCOM, or capsule communicator, said. The LM slowly began moving away from the CSM, leaving Michael Collins alone to contemplate his fate should his colleagues not return from the lunar surface. "Roger, understand," Aldrin responded. Armstrong, clearly in control of the situation, informed mission control, "The Eagle has wings." As the altitude of the lunar module decreased towards the surface of the Moon, an alarm sounded harshly over the loop. "It's a 1202," Armstrong said, reporting the alarm before asking, "Give us a reading on the 1202." Guidance Officer Steve Bales called "Stand by Flight" to Flight Controller Gene Kranz. The 1202 alarm indicated a problem with the LM onboard computers — computers that were guiding the first humans to the lunar surface. Even today, computers have limitations of memory and speed. Modern cell phones have much greater capability than the computers on board the Apollo LM, yet everyone understands the problems associated with running out of memory or trying to use applications that demand more from a processor than it is able to give.

Kranz recalled that moment, "It is like coming to a fork in the road where you're uncertain which direction to take." One can only imagine what Neil, Buzz and Mike were thinking as the descent continued under the guidance of the computer. Indeed, some may wonder whether it would have been more prudent to abort the attempted landing and figure out what was going on with the computer. Within 30 seconds, which seemed like a lifetime to the team in mission control and the astronauts orbiting the Moon, Bales responded, "We're go on that alarm, Flight." There was no pause, no doubt, no questions asking, "Are you sure?" from the flight director, the team in mission control or the astronauts. Rather, the quick response was accepted and trusted by everyone. Where did that trust come from? Why would the astronauts, the flight director and mission control team not question the call? There is only one word which can be used to adequately describe why everyone continued with the landing — competency. The flight control team were living the Gene Kranz credo of tough and competent.

In the hundreds of hours of simulation in preparation for the lunar landing, Gene Kranz had demanded that everyone bring their best; it was a commitment to excellence and unrelenting competency from each member of the mission control team. His leadership paid off. Duke said over the loop, "It's the same one we had in training." The rigorous simulations had prepared them. They were ready. No one questioned the call. Everyone embodied "tough and competent" to ensure that this was the day that NASA would achieve the seemingly impossible goal of landing humans on the Moon and returning them safely to Earth within nine years of flying the first humans in space.

The alarm was the first test of the team. Shortly after receiving the "Go for landing" call from mission control there was a new 1201 alarm. The memory and performance of the computer were overtasked managing all of the data and calculations necessary to successfully land on the lunar surface. The response remained the same: "We're go, Flight." This decision was made with the confidence that comes from having learned from a similar situation during training. As the tension in mission control began to rise with the frequency of the alarms, another more critical problem emerged that might result in disaster, in complete mission failure, with the LM crash-landing on a lunar terrain too rocky and cratered for a safe landing.

To avoid catastrophe the landing would have to be extended. Armstrong calmly took over manual control of the LM and, with the hopes of millions riding on his skill, he began to extend the landing towards safety. As the LM continued to descend, the astronauts as well as the team in mission control tracked the fuel remaining with a vigilance linked to the knowledge that if they ran out of fuel they would crash on the lunar surface. The crash site would be a permanent reminder for the rest of eternity that the first attempt to land humans on the Moon had failed. The Commander's window of the LM remained filled with images of a rocky, cratered lunar surface unfit for landing. Everyone was beginning to think that the landing attempt should be aborted. Everyone except Neil Armstrong, who was now manually guiding the LM on a trajectory to a safe landing site he could see in the distance, past the craters below. Hand-flying a spacecraft known for its instability, one of the best pilots in the astronaut office took control to once again save the day. While the LM represented the pinnacle of technology of the decade, it was a human hand that would determine if the landing would occur safely on the lunar surface.

Neil continued to extend the landing trajectory. The call, "low level" was made on the flight director's loop indicating two minutes of fuel remaining. Buzz was shifting his attention from calling out the altitude and forward motion of the landing module, to assessing the remaining fuel quantity, to watch in awe as Neil continued the landing. In the distance, a flat area suitable for landing emerged on the lunar surface. Duke reported the amount of fuel remaining to the crew, "sixty seconds," in a calm but terse voice. For the millions of viewers watching, it was hard to understand the drama that was unfolding in mission control. Aldrin continued his verbal reports of the LM's descent profile continued: "Down two and a half. Forward. Forward. Good." The fuel was rapidly diminishing. Neil continued to extend the landing, holding the ungainly LM in its landing profile, trying to conserve fuel in his attempt to land successfully. The abort switch, an ever-present opportunity to survive, but also to fail, loomed in front of the crew.

"Forty feet, down two and a half. Thirty feet, two and a half down. Faint shadow," — Aldrin's calls continued as the LM descended, kicking up dust from the only available landing site in the area. "Contact." There was a brief period of silence followed by the words "Houston, the Eagle has landed." They did it! The crew of Apollo 11 and the team in mission control had overcome adversity and achieved the impossible. With an emotion-filled voice, Duke congratulated the crew, "Roger, Tranquility. We copy you on the ground. You got a bunch of guys about to turn blue. We're breathing again. Thanks a lot." With smiles and cheers resounding in mission control, handshakes were being traded, photographs taken. This was history in the making. In the final stages of landing, no one knew if the remaining seconds would end in disaster or in success. The odds were huge, the stress defied description; the illumination of the contact light defined the moment. The blink of a light bulb confirmed success. They had done it. Gene, beaming at the team, reminded them that the mission wasn't over. It was time to get back on the consoles to help the crew get ready for the first lunar spacewalk in history and, later, the moment everyone dreaded, liftoff from the lunar surface to return the crew to lunar orbit and rendezvous with the command module. But for one brief instant that would remain frozen in time in the minds of that remarkable team, they had *achieved the impossible.*

The story of Apollo 11 is the story of teamwork, of leadership, of courage and commitment. It is a story based on a culture undaunted by the challenges of space exploration. It is a story we can all learn from. Ultimately, it defined the NASA culture. A culture that was forged on the experiences of courageous astronauts, engineers, mission controllers and leaders who managed incredible risks to achieve success. It is a story of passion, of excellence, of resilience and learning. It created a legacy that would carry forward through the Skylab space station missions and the Apollo-Soyuz collaborative mission of the next decade into the era of the space shuttle and the International Space Station.

There have been many leadership moments in NASA's history, and every such moment provides a story that can apply to each of us. Whether we are growing as leaders trying to manage the many challenges of working together in organizations large or small, or whether we are interested in learning how to work together on peak-performing teams, the lessons from space are as relevant on Earth today as they have been in the past. Anything worth doing is typically difficult, frequently requiring teams to manage complex challenges to make data-driven decisions that will achieve success. The NASA culture defies the limits of traditional leadership by creating peak-performing teams where each individual strives relentlessly to build their competency, to build trust and to create strong links in a chain that binds the team together. There is an opportunity for all of us to learn from the experiences of the pioneers who began the conquest of space, the final frontier.

CHAPTER 1

The Sound That Changed the World

"Listen now for the sound that forevermore
separates the old from the new."
— NBC RADIO

Of all the sounds associated with the 20th century, it is surprising that a repetitive beep, beep on the radio would change the course of history. The "beep" was first heard Friday, October 4, 1957, originating from the Russian Sputnik satellite that was launched at 10:29 p.m. Moscow time. Later that evening in North America, the first recordings of the distinctive sound from the satellite were heard. The CBS News special that covered the Sputnik launch and its impact opened with an 18-second recorded transmission of the beep, with anchor Douglas Edwards commenting, "That sound had never been heard on this Earth. Suddenly, it has become as much a part of the 20th century life as the whirr of your vacuum cleaner."[1] The sequence of a repetitive 300 millisecond tone, followed by an equal pause, was perceived as a challenge to U.S. technology and the western way of life. It was a sound that changed the world forever.

Seven years earlier, a number of space scientists had met in James Van Allen's living room to discuss opportunities for international

1

collaboration in space research as a part of the first international geophysical year (IGY). A proposal was made to the International Council of Scientific Unions and the announcement was made in 1952 to declare the period from July 1957 to December 1958 the IGY. In July 1955, the White House announced its intention to launch a satellite as one element of its national contribution to the IGY.[2] The announcement was met with an immediate response by the Soviets, with scientist Leonid Sedov sharing his country's intention to launch a satellite in the near future at the Sixth .Congress of the International Astronautical Federation in Copenhagen four days after the American announcement.[3] The growing cold war had set the stage for what would ultimately become the race for space.

Two months before White House Press Secretary James Hagerty announced the U.S. plans for the IGY, President Dwight Eisenhower approved a new space policy that included the launch of a scientific satellite, as well as a military program for the Army, Air Force and Navy to develop launch vehicles. Between 1955 and the beginning of the IGY there were essentially four different programs working to establish a U.S. presence in space with little collaboration between them. The Soviets took a more pragmatic approach that focused on a unified program to combine military launch capabilities with the integration of scientific payloads. By January 1957, Eisenhower was becoming very concerned with the rising costs of the combined programs, which had grown from estimates of $20 million to $80 million, and he agreed to schedule the first launch attempt in October that year using the Navy-led Vanguard space launch vehicle to carry a small scientific satellite.

At the same time, the Army launch capability was being developed by the Army's Ballistic Missile Agency (ABMA) under the leadership of Wernher von Braun, the head of the German rocket scientists brought to the U.S. after the war. While the Navy beat the Army in their 1957 football game, the Army's rocket program was far ahead of the Navy's Vanguard program. In September 1956, a year before Sputnik, the Army had successfully launched a suborbital multistage spacecraft that reached an altitude of 1097 kilometers (681 miles). The Jupiter-C rocket could have carried a satellite, but the scientific mission for IGY had been allocated to the Navy Vanguard program. It was a missed opportunity.

There had always been suspicion about allowing the former German rocket scientists to help the Army develop long-range guided missiles. The Army Redstone rocket flights started in 1952 and von Braun had a vision of attaching enough upper stages to boost a satellite into orbit. "With the Redstone, we could do it . . . to launch a satellite, of course," von Braun said. The Germans were forbidden from doing so, but it didn't stop von Braun from dreaming about it. He embodied visionary leadership and widely shared ideas for rotating space stations to have astronauts working in artificial gravity and flights to the moon. He proposed the idea of a human mission to Mars. At the time his ideas seemed like science fiction but within decades they would become scientific facts.

The July 1957 start of the IGY brought with it the possibility of an imminent Russian launch. The administration recognized the possibility that the Soviets would use a successful launch as propaganda in the growing Cold War, and Eisenhower was increasingly frustrated by the lack of progress and the risk of falling behind the Soviet effort. He pushed the Navy program for the launch of the simplest possible satellite at the earliest date possible. As the summer progressed, the administration suspected a Soviet launch was imminent. However, they underestimated what a successful launch, a carefully planned announcement and the effect of the ever-present beep from the Satellite would mean to the rest of the world, the American public and media who were all caught by surprise. Hagerty's comment at a press conference after the Sputnik launch that they never thought the U.S. program "was in a race with the Soviets," fell on deaf ears.

Two months later, the Navy Vanguard 1A satellite and spacecraft were ready for launch. The team had been demoralized by the Sputnik launch but had quickly regrouped to get ready for the December 6 launch attempt. At 11:44:35 a.m. EST the hopes of the team lifted with the roar of ignition. The vehicle rose roughly four feet in the air, the engine lost thrust and the rocket failed to lift off, sinking back to explode on the launch pad. As if to add insult to injury, the nosecone detached, landing free of the rocket with the satellite still beeping.[4] Given a second successful Soviet launch the previous month, the press were gratuitous with their criticism referring to the Vanguard failure as "Kaputnik," "Stayputnik" and "Flopnik."

Failure is a true test of character. It is what you do when you don't succeed that determines if you will ultimately succeed. Undeterred by the criticism, the Jet Propulsion Laboratory and the ABMA launched the Explorer 1 satellite on January 31, 1958, at 10:48 p.m. EST as part of the U.S. participation in the IGY. The JPL satellite was equipped with scientific instrumentation that was used to detect the Van Allen radiation belt, and the ABMA had modified one of its Jupiter-C rockets to carry the satellite to space. From a scientific perspective, the mission was an outstanding contribution to the IGY, but its achievements were overshadowed by the success of the Sputnik missions and the loss of the Vanguard spacecraft. Sputnik's "beep . . . beep . . . beep" signal lasted only 23 days and it was destroyed re-entering the Earth's atmosphere less than 100 days after its launch, but its impact was far greater than the more important scientific discoveries of Explorer 1 satellite.

"No event since Pearl Harbor set off such repercussions in public life." — WALTER A. MCDOUGALL, HISTORIAN

America in the late 1950s was a land of opportunity. The middle class was rapidly growing and was benefitting from new comforts, goods and services, which were said to be a testimony to the benefits of a modern democracy and national strategic priorities. It was inconceivable that the U.S. had fallen behind in the quest for space. America had to respond and respond it did. The Explorer satellite was the immediate response, but the Sputnik effect was so profound it precipitated the creation of NASA, brought a new priority to scientific research and technology development and set the stage for the upcoming decade of space exploration.

Eisenhower and the National Space Council recognized the importance of coordinating the nation's space efforts into a lead agency that would be called the National Aeronautics and Space Administration (NASA). Given the agency's mandate, it was a surprise to some that the first NASA administrator would be a university president with a background in the motion picture industry.

The Right Leader

The is no specific formula for success as a leader. Perhaps that is why there are so many books, seminars and courses dedicated to the topic. Leadership is based upon a breadth of skills that combine technical and behavioral competencies to influence large groups to achieve results. Eisenhower's Science Advisor Jim Killian felt that Thomas Keith Glennan, then president of the Case Institute for Technology in Ohio, would be a perfect fit for the role.

Glennan recalled, "It was early August, I think, because the President signed that bill 29 July. I had a call from Jim. He said, 'Keith, I'd like you to come down to Washington as soon as you can get here. I want to talk with you about the possibility of your becoming involved here. It's possible that the President would like to see you.'" Glennan responded, "All right, Jim."

He arrived later that night and went to Killian's apartment. Killian showed him the bill creating NASA. After scanning it quickly, Glennan said, "Well, it's fraught with difficulties. . . . But I guess it could be made to work. What is it you want of me?" Killian said, "I want you to be the first administrator." Later, Glennan commented, "I don't have any idea why my name was put in [the] nomination, except that Jim knew me, not well, but we had been moving Case pretty well."[5]

Glennan had studied electrical engineering at Yale University. After graduation, he found himself in a series of leadership roles in the sound motion picture industry, and he subsequently became the Director of the U.S. Navy Underwater Sound Laboratory during World War II. After the war, he was an executive with the Ansco Corporation prior to becoming the president of the Case Institute of Technology in Ohio.

As is the case for many humble leaders, Glennan's comments to Killian about his achievements at the Case Institute were a significant understatement of the role he had played transforming the organization. Glennan had faced a number of challenges: "We had a full plate, trying to build or rebuild a new institution. Again, I'm not an academician. I never taught a class a day in my life. I was relying on people, trying to get good people, trying to raise the money to rebuild an institution."[6] In Killian's mind, Glennan was a proven leader with a skill set that would be critical for the new NASA administrator.

> "What is more important, it seems clear to me that we must find a way to exert leadership as we identify more clearly our own goals. By definition, our present practice of reacting to the Soviet actions must doom us to ultimate failure."
> — THOMAS KEITH GLENNAN

The next day, Glennan met with Eisenhower. "I don't think Jim went in with me, but he introduced me to Mr. Eisenhower. . . . He sat me down, told me about the task that was ahead of us, asked me a few questions. I told him I didn't know a damned thing about which end of the rocket you lit! I'd been very busy putting my institution together, and he said: 'I think I want you to become the Administrator. We want to move fast, so I hope you won't keep me waiting.'" Glennan said he would act as fast as he could with his other obligations and, "It was on the nineteenth of August. I was sworn in."[7]

Glennan understood that assembling an experienced leadership team would be critical to succeed. He had been directed to incorporate the National Advisory Committee for Aeronautics (NACA) organization in its entirety into NASA. His first step was to convince Hugh Dryden, the technical director of the NACA, to become his deputy administrator. Glennan recalled the NACA team "idolized Hugh Dryden. If we had not held on to Hugh, I don't think we would have had as rich a cooperation from the NACA people."[8] Within months he brought together the 8,000 NACA employees, an annual budget of $100 million and three major research laboratories — the Langley Aeronautical Laboratory, Ames Aeronautical Laboratory, and the Lewis Flight Propulsion Laboratory — to create the new space agency.

Building Capacity

Glennan's short-term strategy was to build NASA as a preeminent organization with the best engineers and scientists. His long-term strategy was, "to develop a program on our own terms . . . designed to allow us to progress sensibly toward the goal of ultimate leadership in this competition."[9] Consolidating the nation's efforts in space exploration would be critical to success. Within months, Glennan incorporated several groups involved in space exploration from other federal agencies into NASA. He brought part of the Naval Research Laboratory into NASA, leading to the creation of the Goddard Space Flight Center in Greenbelt, Maryland. He incorporated several disparate satellite programs, along with the Air Force and the Department of Defense's (DOD) Advanced Research Projects Agency research program, which had been developing a million-pound-thrust, single-chamber rocket engine.

In December 1958 he brought JPL, a contractor facility operated by the California Institute of Technology (Caltech) in Pasadena, California, into NASA. The last critical strategic decision was to incorporate the ABMA into NASA to create the Marshall Space Flight Center in Huntsville, Alabama. During Glennan's tenure as administrator, he rapidly built the agency's capacity and established a culture of trust within his senior team. Glennan recognized that he needed to focus on broader strategic issues and the relationship with the White House and administration, while Dryden dealt with the planning of project Mercury and building the foundation for the human spaceflight program. Glennan also created the Space Task Group, with Robert R. Gilruth as its director, and added scientist Homer Newell to his team. The diverse backgrounds and different skills of the four leaders were critical to the success of the agency.

To succeed in sending humans to space, NASA had to continue growing rapidly. Glennan recalled, "I didn't come into the government to add to the government payroll. . . . In the 30 months I was there, we went from about 8,000 [or] 8,500 to 18,000 people, and only 1,600 of them were new hires. The rest were transfers from ABMA, JPL, NRL [Naval Research Laboratory], Army Signal Corps and the rest."[10] Developing the new spacecraft that would be used for project Mercury meant rapidly developing new technologies. The new engineers were critical, but more help was needed.

NASA would need help from the private sector. Glennan set up a meeting with aerospace companies to highlight upcoming NASA programs and outlined a competitive request for proposal process that would be used for procurement. He established three independent committees, one to assess management, another to evaluate experience and task performance while the last focused on technical capabilities. The people on one committee did not know who the other teams were. The recommendations of each of the teams were brought to Glennan and Dryden and they chose the winning bid.

For Glennan, safety was critical. There were many known and unknown risks that had to be managed, and he wanted to establish a process to control the expected and unexpected. He established quality control as a NASA priority at a time when statistical quality control was just beginning to be understood. He said, "When you're talking about Mercury and man-rated flights, I didn't think we could be too careful. We found a man. . . . He was a statistical quality control person [and] he knew his business. I gave him free rein. He could look at anything and require anything done to improve reliability, safety, if there were shortcomings in the manufacturing process, [or] assembly process . . . [he found them]."[11]

Glennan recalled Gilruth coming up from Langley, saying, "What are you doing to us? We don't need this guy. Our life has been made of being careful, being sure. Quality control is a way of life." Glennan responded, "Bob, I'm sorry, you're going to have to do it. I insist that there be an external quality control review on everything that we do."[12] As a leader, Glennan understood the importance of senior leadership commitment in building processes to control risk into the NASA culture.

Gilruth had been a NACA engineer his entire career, and as director of the Space Task group, he understood Glennan's desire to get the best and brightest to work at the agency. Gilruth was relentless in his pursuit of talent. From its earliest days, NASA has been an organization that has attracted outstanding individuals. His quest brought him to Toronto, Canada. It turns out the same day that Sputnik launched, crowds were gathered at the Malton

Airport in Toronto for the unveiling of the AVRO Aircraft Ltd. Supersonic CF-105 Arrow.

AVRO engineer Owen Maynard recalled, "The exact day we rolled our AVRO Arrow out the hangar door for the press to see . . . [there wasn't] great media attention. [We] couldn't figure it out. We had no idea. . . ."[13] Within months the program was canceled, and a number of Canada's top aerospace engineers lost their jobs. To Gilruth, tasked with hiring talented engineers, the announcement was a huge opportunity. He recalled, "I got a call from a friend of mine in Canada who said that the Canadians had lost this big contract for building a fighter. They were going to have to get rid of their top people. Here we were in the Space Task Group just absolutely strapped for top people. . . . It was a big break for us."[14] The NASA team reviewed 400 applications from members of the Arrow team that resulted in hiring 25 experienced aerospace engineers.

By the summer of 1960, Glennan had built a secure foundation for NASA to succeed. It had become the primary agency for all space activities with the exception of a few programs controlled by the Department of Defence. He left NASA in January 1961, a few months before Yuri Gagarin's first flight and Alan Shepard's historic first American spaceflight, to return to his previous position as president of the Case Institute for Technology. He continued in that role until 1966, helping to negotiate the merger of the Institute with Western Reserve University to create the Case Western Reserve University. Yet it was in his role as the founding NASA administrator that he left his legacy by positioning NASA to succeed in the race to space.

NASA was fortunate to have Glennan as its first administrator. NASA historian Roger Launius described him as "the perfect choice to lead the new organization. An engineer who had worked in government, industry and academia . . . Glennan positioned NASA so that it could serve as a vehicle for competing against the Soviet Union in the developing space race."[15] His proven track record as a leader enabled him to succeed, he knew what to do and he did what needed to be done. He built a team that would go on to successfully land humans on the Moon and established a NASA culture that was willing and able to tackle seemingly impossible challenges.

Leadership Insights

- Trust can take years to build when starting from a position of distrust. If leaders are trusted to lead, their leadership should be trusted.
- Learn from the experiences of others. Consolidate where possible and collaborate whenever possible to build capacity, efficiency and expertise.
- Find the best, hire the best, train the best to be better and let them do their job.
- Build a senior team with diverse but complementary skillsets, proven leadership expertise, trust them and listen to them.

CHAPTER 2

Inventing the Future

"The best way to predict the future is to invent it."
—— ALAN KAY

Building the capacity to safely send humans to space takes time
and leadership. It was clear that the Soviet and American satellite
launches were the beginning of a new era of spaceflight. The critical
question was, how long would it take before humans were ready
to travel in space? Shortly after the Sputnik launch, the famous Air
Force pilot and NACA main committee chair James H. Doolittle
asked Guyford Stever, the Massachusetts Institute of Technology
(MIT) associate dean of engineering, to lead a special committee to
assess future needs for space technology. The committee was made up
of many leading space experts including Wernher von Braun, Hugh
Dryden, Abe Silverstein, Bob Gilruth and Dr. Randolph Lovelace, all
of whom became notable for the key roles they would play in project
Mercury. Their report set the stage for human spaceflight, focusing
the first missions to achieve: "The early successful flight of man, with
all his capabilities, into space and his safe return to Earth."[1] When
Glennan created the Man-In-Space-Soonest program in January
1959, he designated Gilruth the director of Project Mercury.

At that point, Gilruth had been leading Glennan's new Space Task Group (STG) for a few months. When asked about the choice of a name for his team, Gilruth said, "What we needed was a good name, and we didn't like to sound like we were too big for our britches. We were a Task Group. We had a big task, and we were a group."[2] He had assembled a great team and a plan, but he and his team had to figure out what would be needed to send humans to space. The program required a human-rated rocket or launch vehicle, a spacecraft, astronauts and some form of control center, most of which did not exist. They had to invent it.

The team was made up of a number of outstanding NACA engineers, including George Low and Maxime Faget. It was the beginning of a group of technical experts that would work together during the next decade to take humans to the moon. Low was a visionary willing to share his bold goals that would shape NASA's future. Gilruth immediately recognized Low's capabilities, noting, "He was good at everything. He was a top-notch engineer and scientist. He could handle people well. He was good with money matters. And he was very easy to get along with. . . . He was an ideal administrator and friend and a big help to us, especially in the days when we didn't have many people. He was worth about ten men."[3] The team worked extremely hard, putting in long hours to do what it took to come up with a plan for human spaceflight. Their report resulted in the creation of Project Mercury, and a three-year journey to launch the first American to space. When presented with the panel's findings, Glennan's response summarized the prevailing sentiment: "Get the hell to work."

Gilruth wanted Low to join the STG to work as the deputy to chief engineer Faget on spacecraft design, to design the Mercury spacecraft. Unfortunately, the team would have to proceed without him. Silverstein also recognized Low's potential and asked him to come to NASA headquarters in Washington to help work on NASA's long-range plans.[4]

Silverstein's request would prove pivotal when Low testified before the Senate Aeronautical and Space Sciences Committee in support of the 1960 NASA proposed budget. Looking directly at the committee chair, Lyndon B. Johnson, Low spoke about the importance of funding space exploration to spur further technology development. "Although Project Mercury constitutes a logical and perhaps only the first step in our manned exploration of space, we are already studying more advanced systems. . . . We will develop the capability of sending several men into orbit for longer periods of time than one day. We will then be in a position to conduct scientific experiments in a series of manned orbiting laboratories. Concurrently, we will make efforts to fly further away from Earth, perhaps to the vicinity of the moon; but a landing on the moon must await the development of boosters that are nearly twenty times as powerful as those that are available today."[5]

Ever the visionary, Low felt that sending humans to the moon would provide a goal beyond the Mercury program, drive the development of new technology and provide the greatest national benefit from the huge cost and risk of human spaceflight. The plan for Apollo would come, but to fly to the moon, the NASA team first had to learn how humans lived and worked in space. Those lessons would come from Project Mercury.

Create an Environment of Trust

Gilruth's immediate task was to ensure the STG had the capacity to do everything that needed to be done as quickly as possible. He clearly understood his team was expected to put humans in space first and bring them back in good shape, before the Soviets did. To get ready for Mercury, the original 35-member STG would grow rapidly, ultimately including roughly 750 people by the end of 1961.[6] The initial group included the Canadians Jim Chamberlin, Owen Maynard and Tecwyn Roberts who had worked together at

AVRO, along with Americans Chris Kraft, Glynn Lunney, Gene Kranz and other experts who had worked together at the NACA.

> "I never remember Bob Gilruth telling anybody what they should do, or how to do anything. He just talked with them long enough that they thought his idea was their idea, and they went and did it the way he wanted it." — KENNETH KLEINKNECHT

The work schedule was the most intense that Gilruth and his colleagues had ever experienced. Quiet by nature, Gilruth did not impose his ideas on the team. Rather, he would ask probing questions that challenged the team to think about a problem from a different perspective. He created an environment of trust that motivated the group to do their best. Duane Catterson, one of Gilruth's team, described him as "an inspirational leader, a gentleman, somebody who just didn't have a mean bone in his body, and yet commanded instant respect because he had such an all-encompassing knowledge of his field and what he was doing, and a great capacity to inspire confidence in the people around him."[7] It would take all of Gilruth's leadership skills, and the team's combined technical expertise, to fly the first humans in space.

Ironically, the first big challenge facing the former NACA engineers was the technical requirements of the Mercury spacecraft. An important yet mundane aspect of this process was how to write a request for proposal, or RFP, a standard step to solicit government contracts. "We'd never written an RFP in our lives," Kraft said. Gilruth had made him part of the core team and his first assignment was to develop a plan for the mission.

"We didn't know a damn about putting a man into space," Kraft said. "We had no idea how much it should or would cost. And at best, we were engineers trained to do [things], not business experts trained to manage [people]."[8] As Kraft juggled the challenge of developing standards for spaceflight — items such as flight plans, timelines, procedures, mission rules, communications networks and contingency management — he found himself thrust into project management. Along with learning to write a government RFP, he was tasked with selecting the aerospace contractors that would work on the program. According to Kraft, the Langley Research Center group considered themselves the "do-it-yourself experts of aeronautics. When we needed something, we requisitioned it, bought it or built it ourselves."[9] That approach had to change.

Current business leaders speak of organizational agility, which is the ability of an organization to respond, adapt and change quickly to succeed in a rapidly changing, ambiguous environment.[10] This agility was also essential during the origins of human spaceflight. The Mercury team knew that they needed "to put an American into orbit around the world."[11] Yet there were challenges in fulfilling that goal. They understood how to design the spacecraft capsule, but there was considerable uncertainty about the organizational and operational infrastructure needed to successfully send and return humans from space. Gilruth let others in NASA engage in the fiery debates that captured the attention of the public and journalists, but in the background, he was the one moving the pieces. Kraft said, "He sat to one side, listened, absorbed, offered a quiet comment or a whispered observation, and not many even noticed that he was there."[12] With Gilruth's leadership, the team figured it out in three short years. By early 1961, it looked like the Mercury program was perfectly poised to launch the first human to space.

The new year started by bringing a cold front to Washington, D.C., that had dropped south from the Great Lakes, resulting in a huge snowstorm the day before newly elected President Kennedy's inauguration. It was dubbed "The Kennedy Inaugural Snowstorm."[13] New winds of change were blowing into NASA as well. Glennan returned to the Case Institute, and James Webb became the new administrator. The new president appointed Jerome Wiesner from MIT as the head of his Science Advisory Committee, and Wiesner decided he was going to investigate the man-in-space program. The first astronaut had been selected, trained and was primed for flight. But a fateful management decision let the Soviets take their place in history.

Gilruth recalled Wiesner's committee: "[They] came to investigate us, thought it [launching so soon] was a bad idea, and [said we] were going to [have to] make sure it was the proper thing for America to be doing. It was a very serious situation. We could have flown Al Shepard before the Soviet manned satellite."[14] The committee's voice carried the day, despite Webb trying to battle the consensus with the White House.

> "I obviously appreciate[d] his [Gilruth's] decision to let me make the first flight, but he never told me why he made that decision the way he did. I asked him several times over the years, and he always said, 'Well, you were just the right man at the right time.'" — ALAN SHEPARD

Unfortunately, the delays impacted Shepard's launch date, and Yuri Gagarin became the first human in space on April 12, 1961. Shepard launched only 23 days later, becoming the first American. Shepard's Freedom 7 Mercury flight captured the hearts of all Americans. Many had felt that the U.S. program was "dead and out"[15] but the response was overwhelmingly positive, with ticker tape parades and nationwide front-page coverage. Although they were behind the Soviets, NASA was still in the game.

Leadership Insights

- Relentlessly ask questions to clarify why and how things need to be done.
- If you know what to do, do it. If you don't know what to do, use the best available data and the opinions of trusted experts to make the best decision possible.
- Organizational agility is based on innovation, teamwork, judgement, skills and knowledge. Build competency to build capacity.

CHAPTER 3

Taking On the Impossible

"The exploration of space will go ahead,
whether we join in it or not, and . . . no nation
which expects to be the leader of other nations
can expect to stay behind in the race for space."

— JOHN F. KENNEDY

How do you build on the success of a single mission in space?

NASA knew that it was possible to bring humans into space. Alan Shepard had captured the heart of the nation with his 15-minute suborbital "hop" — a flight that didn't fully circle the Earth — on May 5, 1961. It was days later that President John F. Kennedy declared that the United States would land people on the Moon before the end of the decade. Unlike President Eisenhower, Kennedy and his team embraced Low's recommendation from the previous year to expand the program and began funding a lunar landing mission.

From that point the pace at NASA escalated quickly. The U.S. had a grand total of 15 minutes of "lived experience" in space. Sure, the astronauts were able — they were former military test pilots. Yes, the engineers working in mission control were smart, often with military or other government agency experience helping their work.

But to get humans from barely above the Earth to a two-week mission landing on the Moon appeared be an almost impossible challenge. This would require a whole new thinking about mission control. Leading the way was Chris Kraft, who created mission control, and Bob Gilruth, who created the Manned Space Center (today's NASA Johnson Space Center in Houston). To get these systems ready quickly, efficiently and safely might have prompted the cry, "Houston, we have a problem" more than a decade before the astronauts on the stricken Apollo 13 spacecraft uttered those words. But Gilruth, Kraft and their employees (mostly made up of men in those days) were always solutions-oriented and, fundamentally, people of action.

How did they set priorities? How did they resolve differences? How did they decide what to do when no one had done it before? Answering these questions took years of work, with some stumbles along the way. Ultimately, the goal was reached. The Apollo 11 astronauts landed safely on the Moon within six months of Kennedy's deadline. But the path to the Moon started inauspiciously on a six-square-mile island off the coast of Virginia.

Wallops Missile Development

As Gilruth recalled in a short memoir entitled *From Wallops Island to Project Mercury*, most of the team building the Mercury program were veterans of the older National Advisory Committee for Aeronautics (which was dissolved to make way for NASA). A key maturing process for the NACA engineers was operating a missile range at Wallops Island.

"Much of the early work at Wallops Island and in NACA was done in support of the ballistic missile program in the United States,"[1] Gilruth wrote. "Had it not been for the ballistic missile development effort, we would not have had the knowledge of reentry bodies, guidance systems or other factors — such as launch rockets themselves — that were able to make possible manned flight in space in such a brief span of time after the space age arrived."

They had some experience with technology, but there were also the leadership principles to consider. While missions were underway, senior leaders were building an organization, building infrastructure and trying to invent the future, all at the same time.

From its earliest days NASA was committed to **hiring the best**. Fortunately, NASA had the funding and plenty of support in Congress as everyone galvanized each other to "beat the Russians" in space. The space race was in full heat. Shepard had done a short suborbital flight into space, while the Soviet cosmonaut Yuri Gagarin had made three orbits and spent several hours in space. Yes, both men had been in this environment — but clearly, Gagarin and the Soviets were ahead.

The schedule pressures were almost overwhelming. "All of our people worked holidays, evenings and weekends. We even worked on New Year's Day . . . but we did take off New Year's Eve,"[2] Gilruth wrote. "Those were days of the most intensive and dedicated work by a group of people that I have ever experienced."

Gilruth also borrowed experts from other parts of government, particularly from the military. Both the Air Force and the Navy loaned flight surgeons to manage the health of the astronauts. Some of them had also participated in creating selection standards for the astronauts, and the famous Randolph Lovelace (who headed the Lovelace Clinic where early astronaut selections took place) agreed to head an advisory committee studying space medical problems.

Delegate

There was so much work to be done that delegation was critical. One of the first people that Kraft tasked was Gene Kranz, who went on to become a lead flight director to oversee mission controllers through the Apollo years.

"I think the real turnaround came about two weeks after I'd been on board,"[3] Kranz recalled in an oral history with NASA. "I knew nothing about the space business, and Kraft came up and he said, 'We're getting ready to launch the Mercury-Redstone 1. I want you to go down to the Cape [Canaveral] and write me a countdown and some mission rules. When that job's done, give me a call and we'll come down and launch it.'"

Mercury-Redstone was the first uncrewed test of Mercury. NASA needed to get the launch procedures figured out before putting an astronaut on board. Kranz's response was laconic, when faced with this daunting task: "I said, 'Uh, uh, uh, okay.'"

His military training began to kick in. Kranz was a Korean War veteran who flew the F-86 Sabre aircraft on behalf of the U.S. Air Force. Almost instinctively, one would imagine, Kranz began to pack the essentials for the trip. The airplane ride was anything but cushy — "rickety old Martins and Convairs"[4] — but it was enough to get Kranz and other people on the project between Langley, Virginia, and the Florida coast, where the launches would take place.

Upon Kranz's arrival, he began talking to a Western Electric employee named Paul Johnson. Kranz knew he was out of his depth immediately and that Johnson was the person who was familiar with how the Cape team worked. "He seemed to know everything," Kranz recalled, but perhaps that's because Kranz was watching Johnson do some delegation and listening himself.

As Johnson and Kranz tried to think about launch procedures, they found themselves relying on their teams as the people around them talked things through. "You sort of had to learn to anticipate what everybody was going to ask for, what they were going to do. I had to know the telemetry, the command. I had to know what the mission was about. I had to be able to anticipate everybody's call. So it was interesting sitting back, almost like one of the guys from the press [corps]. You know, writing the story as it's going on, only trying to stay just a few minutes ahead of the story."[5]

Kranz's people also weren't afraid to challenge his expertise. Johnson, for example, introduced him to the concept of a countdown. "I thought, 'Hell, a countdown can't be too difficult. It's just a set of procedures that makes sure everything's ready before you launch something.'"

It didn't take long before Kranz was humbled. Procedures were the easy part he came to realize. Go/no go is an easy call to make when the astronauts are safe and they haven't left the pad. But it's the mission rules after launch that really take guts and trust — when the astronauts are already in a dangerous role. "All of my mission rules talked about what happened before we launched. None of them talked about what happens if things go wrong after we launch," [6] Kranz said.

That first Mercury-Redstone flight did not go to plan, despite the expertise, despite the willingness among team members to delegate and be humbled by learning new things. Called the "four-inch flight," it could have been construed as an embarrassing moment. The engine shut down almost instantly after launch, and the rocket settled back down on the pad.

Knowing-Doing

Then, instead of the rocket launching, the Mercury capsule's escape rocket inadvertently fired, jettisoning itself and the spacecraft about 400 yards in the distance.[7] The Mercury spacecraft dutifully deployed drogue and parachute, so that was good, but there was a fully loaded rocket still on the pad in windy weather. The team was worried about the possibility that it could still launch or perhaps even explode. Kraft rejected calls from some team members to do hasty things to mitigate the risk, such as shooting holes in the propellant tanks with a rifle. Ultimately, they decided to monitor the rocket and let the batteries gradually discharge, which worked.

"If you don't know what to do, don't do nothing,"[8] Kraft famously said of the incident. This incident was also a clear example of what Stanford professors Jeffrey Pfeffer and Robert Sutton later called the "Knowing-Doing Gap." This principle focuses on overcoming the inability to put new ideas into practice.

"One of the main barriers to turning knowledge into action is the tendency to treat talking about something as equivalent to actually doing something about it," the authors wrote in their 1999 book, which is still popular among leaders today.

"Talking about what should be done, writing plans about what the organization should do and collecting and analyzing data [will] help decide what actions to take and can guide and motivate action,"[9] they continued. "Indeed, rhetoric is frequently an essential first step towards taking action. But just talking about what to do isn't enough. Nor is planning for the future enough to produce that future. **Something has to get done, and someone has to do it.**"

In the Mercury program, the new idea was launching something into space. The inability was clear to see in front of the bemused controllers — the rocket had not lifted off. Only the escape system functioned. They were lucky not to have blown themselves to pieces in the process. So, what to do next? How to bridge the gap? As the book's authors suggest, one key concept is gathering the data and information about what needs to be done. If we know what needs to be done, we can do it.

Getting ready goes far beyond planning, to doing — practicing, simulating, imagining what can go wrong and coming up with the right scenarios to address this. "Just as mission statements and talk can substitute for action rather than informing such action, planning can be a ritualistic exercise disconnected from operations and from transforming knowledge into action,"[10] the *Knowing-Doing Gap* authors wrote. "Of course, planning can facilitate developing knowledge and generating action. But it does not invariably do so and often does the opposite."

Failure Can Be Instructive

The team learned quickly and were soon ready for John Glenn to attempt an orbital flight that would go around the entire Earth a number of times. After a few slips of the launch date and delays during the countdown, at 9:47 a.m. local time, on February 20, 1962, liftoff occurred. Five minutes into the mission, Glenn achieved orbital speed, reporting: "Zero G and I feel fine. The capsule is turning around . . . that view is tremendous."[11]

The mission had been planned for three orbits, but it could be extended to seven orbits if necessary. However, a little over four hours into the mission, when Glenn was over Hawaii, he received a call from mission control that changed everything: "Friendship Seven, we have been reading an indication on the ground of a segment 51, which is landing bag deploy. We suggest this is an erroneous signal."[12]

Landing bags weren't supposed to deploy, as the name suggests, until the spacecraft was landing in the ocean. To have a landing bag deploy in space was an emergency, especially since that could mean the attached heat shield was loose. Without a heat shield, the spacecraft could burn up during reentry in the Earth's atmosphere. It was the first emergency in space, and nobody knew exactly how to handle it.

Flight Director Kraft, scanning the readouts from his deputies, said his "gut" told him they were seeing a false signal, especially since Glenn had reported nothing from space. "How could all of that happen without Glenn hearing something?" Kraft asked. "Wouldn't there be loud noises . . . probably some clanging and banging from the heat shield every time the capsule's thrusters fired? And wouldn't Glenn instantly report it if his own instrument panel showed deployment?"[13]

Shepard — working as one of the capsule communicators with Glenn in space — confirmed Kraft's suspicions. The men referred to the contingency procedures while Kranz, working the procedures console, sent alerts to the various communications stations around the world informing them about the strange signal.

Deputy Associate Administrator Walt Williams stepped in. He got senior NASA management involved and rapidly, several people appeared beside Kraft's station in mission control. A decision had to be made before Glenn's reentry, but no one knew what the best solution was. There was also no confirming data showing that the landing bag had deployed. Kraft was sure they were seeing a "spurious signal," and when Shorty Powers, the NASA public affairs person, radioed the problem to the press the reporters recognized that something was amiss. There was a growing sense that they were gripped in a life-or-death drama. "Millions of people stopped to watch and listen," Kraft recalled.[14]

Data-Driven Decisions

The decision time was rapidly approaching, and still nobody in Mercury Control could confirm the signal was false. An idea was circulating to keep the "retrorocket" package — the set of engines that maneuvered the spacecraft in orbit — on after it was supposed to be jettisoned to slow the capsule for reentry. Kraft felt it was a risky decision in the absence of any engineering analysis, particularly if there was residual fuel in the retrorockets that could explode. But if the heat shield was indeed loose, perhaps it could save Glenn's life. Despite personal misgivings, Kraft made the decision to leave it on for reentry. "We'd taken an action that I still considered dangerous, and foolhardy, in leaving that retropack on and letting it burn away," he recalled. "We could have killed John Glenn just as surely as not. We were lucky."[15]

Glenn made it down safely, but it was a close shave. NASA quickly added the capability for more accurate telemetry from the spacecraft, and Kraft created a new mission rule: **"Never depart from the norm unless it is absolutely required."** Kranz recalled adding more data sources after Glenn's mission to help confirm what the mission control team was seeing: "The business of being able to look deeper into the systems really came about as a result of some very difficult decisions we made with inadequate information."[16] They didn't have much time to make changes, however. Scott Carpenter's mission was scheduled in just a few months, and a lot of work had to be done to get ready.

Carpenter's mission also had a major problem, even though the technical side worked out perfectly. "It could have been our best mission yet," Kraft lamented. This time, Kraft recalled, "the man malfunctioned." In retrospect, Carpenter had been overtasked with a very busy experiment schedule including testing out the spacecraft's maneuvering capabilities. He was so busy that he inadvertently activated the highly sensitive control jets early in the mission and used so much of his fuel that he had less than 50% remaining by the end of his first orbit.[17] That was a problem, as Carpenter — like Glenn — was scheduled to fly for three orbits.

Carpenter was able to conserve fuel on the orbit but when he tried to switch to automatic control, he found the system would not hold the required spacecraft attitude. Trying to solve the problem, he inadvertently turned on both the manual and automatic systems for 10 minutes, using even more fuel. As his spacecraft passed out of range over Hawaii, the mission control team requested that he align the spacecraft for reentry. The firing of the retrorockets was three seconds late, causing the spacecraft to overshoot the targeted landing area by 175 miles. The recovery ships raced out to fetch him and an hour and a half after landing, the ships spotted and recovered him from his life raft. Despite the mishaps in space, he was perfectly healthy.[18]

"I'll give him credit for wising up fast," Kraft said of Carpenter's last few minutes in space. The mission underscored the complexity of human spaceflight and the critical relationship between the astronaut and the team in mission control. Rigorous attention to timelines, checklists, procedures and mission rules were as critical in space as they were on the ground.

Gemini

Mercury was a hard lesson learned in trying to gather data for decision making, which would be essential to prepare for the Gemini program. It was coming up fast and the amount of data from Gemini would increase as astronauts began to perform spacewalks, rendezvous and dock with other spacecraft; all activities required to get ready for Moon landings.

"I had a plan . . . had a schedule and I had a check to do it with. I had the country backing me,"[19] Kraft said of the plan to bring astronauts from Earth orbit to the Moon. "I had the Congress backing me. I had the greatest people in the world working for me, and I was working for the greatest people ahead of me. So that's a pretty good formula . . . [but] I don't think we knew what the hell we were accomplishing when we first started. We were all flight test engineers. Gilruth was a flight test engineer, and that's what I was. How we get higher and faster. That's all we were trying to do."

Closing the knowledge gap would require the Mercury team to use the best available data to make the necessary decisions for astronaut safety and mission success as they embarked on the Gemini missions to get ready to go to the Moon.

After John Glenn's fateful Mercury flight, the spacecraft designers had to develop new onboard sensors to get the information they would need to make data-driven not experience-based decisions. Kranz said. "If we believed that measurement and the heat shield had come loose, we had one set of decisions that involved sticking our neck out [by] retaining the retrorocket package attached during the entry phase. We didn't know whether it would damage the heat shield . . . but if the heat shield had not come loose, that measurement was wrong, and we wouldn't do anything different. So, it was a very difficult decision for Kraft in Mercury Control."[20]

They succeeded. The new Gemini spacecraft went from dozens of pieces of data coming down from its predecessor to hundreds of different data streams. The additional insight they would have in Gemini excited Kranz, Kraft and Gilruth. "We could look within the guts of the system and see the things that the astronauts couldn't see on board the spacecraft,"[21] Kranz recalled. "We had data at a much higher sample rate; we now went to the point where, instead of getting one sample a second, we'd get eight and ten samples per second."

But data would not be enough, communication would be essential — regular reports within the teams about lessons learned, how to change the spacecraft development between missions, how to make the procedures and debrief points part of the lifeblood of mission control.

With all of this information regularly discussed in meetings, Kranz said, "**We started doing pretty good detective work, looking into the systems, and we were able to stay ahead of the problems and, to a great extent, prevent their occurrence.** Or if they did occur, we were able to very quickly identify the source cause and what we were going to do about it."[22] Both the Mercury and Gemini programs transformed ideas into action.

Developing Talent

The rapid growth associated with running one program while preparing for another created an unprecedented demand for bright young engineers. "Looking at it with the longer perspective several years later and looking at the very rapid growth, not only in the space program but in the need for bringing young people into the program, I believe one of the early decisions to locate [NASA] centers near sources of young people — what you call 'feeder universities' — was really a right decision,"[23] Kranz said.

"Basically, we had a source of young people, and that was really the fuel for this space fire that Kennedy had built . . . by the time we started the search for the raw talent we needed to go to the Moon, this was the right decision because we could go to universities and we'd bring in entire graduating classes."

Consider the ambitious timeline of Mercury — six astronaut launches and landings in two years in a spacecraft so primitive that the earliest versions did not have a window. The astronauts would stay up there, cocooned in their spacesuits, for anywhere up to 34 hours. All of the data was furiously gathered, then discussed, then brought forward to improve future missions. Gemini pushed the program forward at the same furious pace until late 1966, when it was time to get ready for the Apollo lunar missions.

Today's approach to human spaceflight evolved from the lessons learned in the Mercury and Gemini programs. **Teams closed the knowing-doing gap by gathering data, creating milestones for check-ins and implementing what they learned to improve spaceflight.** Their wisdom has withstood the test of time.

"You know, these were the Kennedy years. These were the Camelot years. And for a period of time, you believed that this Camelot really existed. And it did exist! It was just a magical time,"[24] Kranz reflected of that heady era. And what a legacy it left for Gemini and beyond.

Leadership Insights

- Bold visions can inspire teams to achieve bold objectives.
- Hire for expertise, delegate to drive results.
- Create a process to manage new ideas. Implement the best ones, measure their success and put others on standby for when they are needed.

CHAPTER 4

Failure of Imagination

"The inability to predict outliers implies the inability
to predict the course of history."
— NASSIM NICHOLAS TALEB

"We have a fire in the cockpit!"

The announcement of trouble from the Apollo 1 crew disturbed what was otherwise supposed to be a routine pre-launch test. Commander Gus Grissom and his crewmates, Ed White and Roger Chaffee, were sealed in their command module spacecraft on the launch pad. On January 27, 1967, they were weeks away from launch and from pushing the new Apollo program towards Moon flights in earnest.

From that first distress call, the astronauts — all trained test pilots in their previous career — sprung into action. White and Grissom began unlocking the hatch. Compared to previous spacecraft, the new design was a complicated process that required turning several gears and opening the inward-facing, heavy structure. Chaffee, as was procedure during a fire emergency, remained seated while awaiting the attempted egress.

The crew was highly trained and they knew exactly what to do but time was against them. A NASA history document notes that only 17 seconds passed between the first voice transmission and the last noise heard from the crew.[1] In an instant, flames covered the entire command module interior. It was a long five minutes before assistants at the launch pad were able to open the hatch to get inside.

It was too late. All three crew members had asphyxiated. The loss of the Apollo 1 crew was a tragedy beyond measure, a catastrophe to the families, to the entire NASA team, and to the technicians on site who were unable to save the three young men. It also was a tragedy that almost ended U.S. human spaceflight, less than six years after Alan Shepard made his historic flight in May 1961.

Fire in the Cockpit

In retrospect, some have questioned the idea of filling a spacecraft with 100 percent oxygen above sea level pressure, with a number of flammable materials inside and then closing the hatch — leaving a hazard so great that a single spark would be enough to create a flash fire. The cause of the fire was unclear, although shoddy wiring underneath the hatch was the most likely location for the spark.

In any accident investigation, there are many things to consider. Did the aggressive schedule contribute? NASA was supposed to have people on the Moon in less than three years. Had the dangers of flash fire in an oxygenated environment been overlooked? Years before, a Russian cosmonaut had died in a training accident in a pure oxygen environment, but it is unlikely that the NASA team was aware of the incident. And why had Grissom, a veteran of two spaceflights, been so frustrated with the performance of the spacecraft?

The investigators had to look at everything that might have contributed — by reviewing the readiness processes, the spacecraft design and build, the physical factors and the decision making leading up to the event. Even if the physical cause of an accident is identified quickly, the investigation and recovery is a lengthy process. The program was on hold, but critical time was passing as NASA mourned, recovered and shared its findings with Congress.

At NASA, the schedule pressure was creating issues among the teams. Last-minute ideas were passed between the manufacturer, NASA and subcontractors with no clear process to track changes. A new spacecraft builder — North American Aviation — was taking over and there was growing frustration with the lack of progress.

Balancing Risk and Readiness

On January 22, 1967 — five days before the fatal fire — Grissom dramatically shared his frustration with the command module. He was making what turned out to be his final visit home in Houston and grabbed a lemon from one of the trees in his backyard. When his wife Betty asked what he planned to do with it, Grissom's response was: "I'm going to hang it on that spacecraft."[2] Some accounts said he hung it on the spacecraft itself at the Cape, while others say it was a simulator. It would take the tragic loss of the crew for things to change.

Balancing the pressure to meet politically dictated schedules with operational readiness is what doomed the Apollo 1 astronauts and other spacecraft and crews.

The Apollo 1 investigation board enumerated many causes, but one of them was what we would today call "culture." "Problems of program management and relationships between [the] centers and with the contractor have led in some cases to insufficient response to changing program requirements,"[3] the board found in its report. They called for "maximum clarification and understanding of the responsibilities of all the organizations involved, the objective being a fully coordinated and efficient program."

Did team members speak up about the issues? Were they listened to? "We have a whole team of people, including representatives of our flight safety people,"[4] Apollo 1 lead investigator and astronaut Frank Borman told Congress, about preparations for a typical flight. "In preparing for Gemini 8, I had sixteen people that reported directly to me . . . who I used as my eyes, ears and bird dogs for making sure that the things were going the way I thought they should go."

But when asked whether this process could have been improved for Apollo, Borman deferred the answer. "I would hesitate to answer this offhand; I haven't thought about it until you asked the question. Perhaps I could defer and answer this later on for you."[5]

After the fire George Low was brought in to lead the Apollo team first to return to flight and then to go to the Moon. His leadership style **encouraged team members to identify safety issues; to speak up and bring their ideas forward to improve the mission.** His approach would later be described as the leadership attributes needed to run high reliability organizations. For Apollo, it was what was needed to get back on track.

Black Swans

The full hatch-closed test was clearly a planned activity, but the readiness review process, designed to ensure the team is ready to do what they plan on doing, did not identify the flammability hazard. In pre-launch readiness polls any systems failure or anomaly would be enough to scrub the launch. The same rigor in determining readiness for ground testing is as critical to the success of the test as it is for a launch.

The plugs-out test had been approved — but not all items were included in the review. The investigatory board noted that a major revision to the procedure was issued at 5:30 p.m. EST on January 26, 1967, about 24 hours before the test and four additional pages for the procedure were issued the morning of the test.[6] Last minute changes to test procedures may increase risk and the lessons from

Mercury and Gemini reinforced the importance of developing protocols and where possible following them without deviation.

Cascading problems create frustration that can lead to cascading errors. When the Apollo 1 spacecraft was delivered to Cape Canaveral there were a number of problems that had to be resolved to get it ready for the test that included deficiencies in the wiring, coolant leakages, failures in the life support system and problems with the radios.

Grissom shared his frustration openly on the communications loop, asking the test team, "How are we going to get to the Moon if we can't talk between two or three buildings?" There were numerous delays during the simulated countdown and when the hatch was finally secured at the end of the tiresome day the countdown resumed and the spacecraft was pressurized with pure oxygen. This had been done many times without incident in the Mercury and Gemini program and perhaps that experience had created a false sense of security about the hazard.

Some reports say Borman later called the Apollo 1 fire a "failure of imagination" in trying to think about what could go wrong in an otherwise routine event, although this may be an apocryphal expression. In 2007, business author Nassim Nicholas Taleb referred to such extremely rare events as "black swans." "The inability to predict outliers implies the inability to predict the course of history,"[7] he wrote in his 2007 book, *The Black Swan*.

NASA did not appreciate the cascading effect of the series of events that resulted in losing a crew and spacecraft and the consequences it would have on the program. Now the agency had to rebuild trust with the astronauts, the public and Congress. Low's competency as an engineer and leader helped.

One of the ways public confidence was restored was through Borman's testimony at the Congressional hearings. He was candid and truthful with his answers while remaining loyal to the Apollo teams and the agency. His responses to individual representatives were critical in rebuilding trust. If somebody questioned him about how astronauts would feel, Borman would bring up his own flight experience — 14 days flying in a pure-oxygen Gemini spacecraft, he said at one point, gave him confidence in having an oxygenated environment for Apollo in space as well. The congresspeople praised Borman and his fellow investigation board members for their attention to procedures in the investigation.

Borman also knew to defer to authority, when he had to. One congressperson asked him point blank if it was "proper to accept the spacecraft"[8] for testing in its half-finished state. Borman, ever the military man, referred the answer up the hierarchy. "This is an area you should discuss with the program management and the people responsible for accepting the spacecraft," he answered. When Borman was asked if his own boss, associate administrator George Mueller, was aware the spacecraft orders were not finished, Borman refused to speak for him. "You ought to defer that for Dr. Mueller to answer," Borman said.

Commitment, Courage and Resilience

In a 1999 oral interview with NASA, Borman recalled that all the astronauts knew about the problems with the Apollo spacecraft, attributing it to a new, complicated vehicle. But nobody felt, until the fire, that these problems would end up being fatal. His reaction, upon crawling into the empty, burned Apollo 1 spacecraft for a first investigation, was "I can't believe it could happen." He told NASA he and his colleagues were focused on documenting what went wrong and learning what needed to be fixed.

"I had a job to do, and my Job [Number] One was recording the switches. Then the next thing was, we went through and tried to understand where there might be bad insulation on it. So it was a long, drawn-out process."

When asked if Borman ever thought the agency could fail, that it would never bring a human up in space again, his answer was just as forthright as what he told the congresspeople. He was not afraid of that, he explained. "I guess it's just the inherent optimism that people have that, you know, we stubbed our toes," Borman explained, then recalled the fate of some of his fellow test pilots during his days in the U.S. Air Force. "And you've got to remember: Look, I've been around quite a few people that made holes in the ground. And that's it. You press on."

> "The conquest of space is worth the risk of life."
> — ASTRONAUT VIRGIL I. GRISSOM

NASA pressed on, although it wasn't easy. The spacecraft hatch was redesigned. The wiring was inspected for deficiencies. Thousands of changes were implemented. Low described the changes his team made: "We once made a serious mistake, a mistake of not maintaining absolute control over all flammable material. Since then, we have made every conceivable effort to avoid similar mistakes. We have re-examined every drawing, every circuit and every component of the Apollo spacecraft. We have made thousands of changes in design, in manufacturing and in tests."

With Low's leadership, the teams made a commitment to improve communication and to listen to the astronauts. The astronauts had said they were tired of the "not invented here" syndrome they were hearing in recent conversations with North American Aviation and its contractors. The tragedy renewed everyone's commitment to safety. Issues were considered, managed and resolved. Schedules were managed, the pace was still intense but Low's willingness to listen and act meant issues did not become bigger problems.

Twenty-one months after the fire, the Apollo 7 crew successfully completed an 11-day mission to test the new spacecraft systems as it orbited the Earth. The renewed attention to safety while controlling the risks associated with testing the redesigned command and service module had worked. The mission was a big step forward on the path to the Moon.

Grissom had been unsuccessful in getting the changes he wanted in the Apollo 1 spacecraft. His comments went unheard. That changed after the fire. The Apollo 7 commander Walter Schirra stuck up for his crew on Apollo 7 during training and in space, occasionally canceling things when necessary to reduce the workload. Managing the relationship between the crew and mission control team was hard but it was critical to the success of the program. Apollo 8 commander Borman learned from that experience and while he was a soft-spoken straight shooter, he was firm when talking about mission priorities.

"I think the worst fear that I had was that somehow the crew would foul up, and that was the one thing that I did not want to happen," Borman recalled of Apollo 8 in his oral history. "I had a great team in Bill Anders and [Jim] Lovell, and I wanted to make certain that . . . we could handle whatever was handed our way. The second thing was, I didn't . . . really want the mission to get fouled up because we really weren't certain that the Russians weren't breathing down our backs. So I wanted to go on time."

Not only did Borman and his crew "go on time" they succeeded in doing something that had never been done before, traveling safely to the Moon and back. With one year remaining before the end of the decade, NASA was poised to achieve the impossible.

Leadership Insights

- Rebuilding trust takes time. Leaders who work with integrity, truth and transparency combined with a commitment to finding and solving problems play an important role building, sustaining and rebuilding trust.
- Humble leaders who ask questions, listen thoughtfully and encourage team members to speak up can change organizational culture.
- Readiness reviews can help ensure that the team is ready to do what they plan on doing. Identified open issues and uncontrolled risks must be resolved.
- Senior leadership is critical to creating a safety culture.

CHAPTER 5

What Do the Numbers Say?

"Without risk, there can be no progress."
— GEORGE LOW

In the early days of NASA, it was hard to go far in agency circles without bumping into either George Low or somebody who was heavily influenced by him. Low was there from the very beginning. During the early days following the creation of NASA as a civilian space agency in 1958, he was part of the team that helped build the foundation for the Mercury, Gemini and Apollo programs. Despite his desire to be on the front lines of designing and building these spacecraft, Low moved on to the agency's new headquarters in Washington, D.C., to become chief of manned spaceflight.

Low was the penultimate team player and readily accepted his separation from Johnson Space Center, the newly built hub of human spaceflight in Houston where the astronauts lived, worked and trained for their spaceflights, and Cape Canaveral in Florida where the launches took place. He recognized it was a necessary move. By being in Washington, Low was highly accessible to congresspeople.

He was near the crucial decision-makers that kept the money flowing to NASA.

And he was one of the early voices pushing for NASA to go to the Moon, long before President John F. Kennedy made that fateful announcement in 1961. So vocal was Low that one journalist is said to have called him "the original Moon zealot." Low, ever the visionary, was calling for a Moon landing at the very early stage of 1959. That was two years before any American flew into space.

Low was a member of the Goett Committee that was asked to create a long-term strategic plan for NASA, which it released in April 1959. Of course, the committee focused its first priority on Project Mercury to put "man in space soonest." They felt that a program was needed beyond Mercury and their vision was ambitious for what the future might bring. In human spaceflight alone, it called for a "maneuverable manned satellite," a "manned space-flight laboratory" and the lunar landing.

Some of the committee members, according to the NASA history book *Chariots for Apollo*, were only willing to commit to a circumlunar mission, Low recalled in the book. "I remember Harry Goett at one time was asked, 'When should we decide on whether or not to land on the Moon? And how will we land on the Moon?' And Harry said, 'Well, by that time I'll be retired and I won't have to worry about it.'"[1]

Despite Goett's wry humour, Low wasn't ready to retire, either in the literal sense or regarding the idea of getting to the Moon safely and efficiently. "Without risk, there can be no progress," Low was often quoted as saying. He was the one who wrote the NASA report recommending that Kennedy commit to landing on the Moon, which Kennedy accepted when he challenged the nation to commit to a lunar mission during his speech at Rice University in 1962. The U.S. president is often the one credited with bringing NASA to the Moon, but in reality, it was on the advice of Low that the program got moving.

Great Leaders Listen

Not everything was clear to the committee in 1961. At the time, there was considerable debate about how to achieve a landing. Most people "in the know" were behind an idea known as Earth orbit rendezvous. That would have the command module and lunar lander fly into space on separate rockets, meet up in Earth orbit and then fly to the Moon together.

Despite his own engineering expertise, as a leader Low valued and sought out the opinions of the expert NASA engineers. He knew the best approach to doing something that had never been done before would emerge through discussion, disagreement and consideration of all recommendations. Low found himself humbled in the early days of planning the lunar mission by a young engineer named John Houbolt. He was leading the team that had come up with another idea to have the command module and lunar lander launch together on a single rocket; the lunar lander would descend to the Moon's surface and return for what was called lunar orbit rendezvous.

It was a nifty idea, with one hitch. At the time Houbolt and his team members were thinking through the idea, nobody had done a rendezvous in space, ever. Not even in Earth orbit. Understandably, NASA was primed to be cautious and to think that Earth rendezvous would be the best solution. But Houbolt had a powerful argument for cost and complexity. With only one rocket to launch, there were fewer potential points of failure. The cost savings would be considerable, too.

Low admitted it took some convincing. "We knew Houbolt was working on it, and occasionally people told us about it, but at first nobody thought it was a worthwhile approach,"[2] Low said in the NASA History book *Before This Decade Is Out . . .*, which recounts the varied opinions of the leaders in the Apollo program.

"This, historically, was the case with everybody who looked at it," Low continued. "We were horrified at the lunar rendezvous approach the first time we saw it, and it was only after we studied it in depth, as Houbolt was doing at the time, that we became convinced that this was really the way to go."

But before NASA could go to the Moon, it had to demonstrate that spacecraft could rendezvous in Earth orbit. Low transferred to Houston in 1964 to help see the Gemini program through its tricky tests to get ready for the Apollo program and the upcoming landings. Successfully rendezvousing with and docking to another spacecraft, when both are travelling 25 times the speed of sound, is not an easy process. The first attempt ended in near disaster when Neil Armstrong and Dave Scott hooked up with an Agena target vehicle in Earth orbit in 1966.

Due to a thruster stuck open on the Gemini 8 capsule, the docked spacecraft began rotating rapidly. The astronauts separated from the target vehicle and found their capsule spinning and apparently uncontrollable at a disorientating rate of one revolution per second. The quick-thinking Armstrong engaged the reentry system to stabilize the spin, ending the mission prematurely but saving the crew's lives. Based on the experience of the Gemini 8 mission, future crews proved the docking idea would work.

Footsteps to the Moon

The other big piece was figuring out how to do extravehicular activities, or spacewalks. Few people understood at the time how to maneuver safely in a spacesuit. During the first U.S. spacewalk, Ed White went outside the Gemini 4 spacecraft to test a manoeuvring device that looked like a wand with compressed gas thrusters at both ends to move from place to place. The device basically worked, but it was determined that greater control was needed and it would be best for the astronauts to move from place to place using handholds. The trouble was the early Gemini missions didn't have handholds available on the exterior of their spacecraft.

On Gemini 9A, Gene Cernan got everyone's attention when he badly overheated while trying to move around the spacecraft towards a docking adapter. On Gemini 10, Michael Collins had a little more success using a tether, but the arrangement still felt awkward and needed improvement. Gilruth, now director of the Johnson Space Center, stepped in telling the team, "I have given a great deal of thought recently to the subject of how best to simulate and train for extravehicular activities. Both zero 'g' trajectories in the KC-135 [aircraft] and underwater simulations should have a definite place in our training programs."[3] Lovell and Aldrin, the crew of Gemini 12, had to find a solution. Buzz Aldrin used the underwater simulations to test the idea of handholds to provide control moving outside the spacecraft. During the mission he confirmed their utility, moving around easily on his spacecraft thanks to the handrails he installed on Gemini and Agena.

These decisions about Earth orbit rendezvous, spacewalking and docking were made by many people, but Low was very effective as a leader working behind the scenes using his engineer training and when necessary making the case for funding and maintaining the mission sequence to the budget hawks in Congress. He led the charge on changes to the Apollo spacecraft after the deadly fire of Apollo 1 as manager of the Apollo Spacecraft Program Office (ASPO) with new safety practices and a new configuration control board that monitored the many technical changes in redesigning the Apollo systems.

For those inside NASA it was Apollo 8 that was particularly memorable as a bold step forward to achieving the goal of putting humans on the moon by the end of the decade. It was a mission whose mandate changed at the last minute. NASA's command module was finally ready and proven in space, but the lunar module was lagging. After the experience of Apollo 1, nobody wanted to rush the LM development. But there was secret information indicating the Soviets were considering their own Moonshot. NASA, whose mandate was heavily built on making the United States the first nation to land humans on the Moon, found itself facing a critical moment.

Bold Decisions

Low took a step back and considered the big picture, looking at what could be achieved with the existing capability. The lunar module, he determined, could delay the lunar landing if they followed the original sequence of flying Apollo 8 in Earth orbit to further test the command module. Although the mission would be the first crewed flight of the Saturn V rocket, it would essentially end up being a repeat of Apollo 7. And with 1968 rapidly waning, the possibility of landing astronauts on the Moon in just over a year might disappear if they waited on the LM to get there.

"Politically, of course, it was a bad [risky] decision,"[4] Low said in his 2019 biography, *The Ultimate Engineer*. "Remember, Apollo 8 came along after the fire, at a time when only two Saturn Vs [Moon rockets] had flown [uncrewed]. The second one had several failures with it, and men had not yet flown in Apollo. So, politically, if you look at it, I'm sure that [NASA Administrator] Jim Webb must have thought we all had lost our minds."

But in the context of June and July 1968, when the LM was sitting at the Kennedy Space Center having trouble completing its certification tests, Low realized the initial launch date of December 1968 could not be assured. No one wanted to rush this spacecraft forward. Perhaps, he reflected, he could use what was available to "take a major step forward in the Apollo program."

Low himself went over the command service module, known as CSM 103, and was assured it was "extremely clean" and ready to go. By August 1968, he had come with the idea of a lunar mission with only the CSM and asked Chris Kraft to think hard about the possibility of flying a lunar orbit mission without the LM. Low also brought the idea to Gilruth to see what his reaction would be. Gilruth was ready to back the idea, and Kraft bought in, too.

Given the risk, Low wanted to ensure that he had everyone's thoughts on the technical and operational challenges of going to the Moon before the end of the year. He met with Deke Slayton (head of the astronaut office) to see what the astronaut office thought, assuring Slayton that the mission was "technically feasible from the point of view of ground control and onboard computer software." Low and his team next went to Huntsville to speak with Saturn V designer Wernher von Braun and Samuel C. Philipps, director of NASA's Apollo manned lunar landing program, and representatives from other centers as well.

It was a challenging time. The Apollo team had been working very hard to recover from the Apollo 1 fire. For 18 months, the parking lot at Johnson Space Center was full on evenings and weekends as everyone did what was required to get the Saturn V rocket, the CSM and LM ready. Despite the schedule demands and relentless work pressures, the motivation to succeed never waivered. Low used the CCB to understand the challenges the team faced; it gave him the insight he needed to trust them to solve the technical challenges they faced. He respected their expertise and gave them the autonomy they needed to do their jobs, jobs that required them to do things that were inherently extremely difficult.

In his book *Drive*, published 40 years after the Apollo lunar landing, author Dan Pink described the forces that motivate teams to achieve greatness.[5] Pink reviewed four decades of research on human motivation leading him to propose three elements that are critical to understanding motivation: "autonomy, mastery and purpose." Each of these were part of Low's leadership style. There was no challenge greater, nothing more difficult to master, than developing spacecraft that would land humans on the Moon and arguably, at the time there was no greater purpose than achieving the president's goal before the decade was out. The achievements of the Apollo program speak to the power of Pink's message and the wisdom of Low's leadership and are an important legacy of NASA's commitment to teamwork.

Nobody at the meeting could come up with a major technical objection to Low's suggestion, so the group adjourned with the promise to think things over and to come to a final decision at a meeting in Washington. They listened to every senior manager they could think of who would be relevant to the decision. The team tried not to succumb to "go fever" as they considered everyone's concerns; even when George Mueller, NASA's associate administrator heading the Office of Manned Space Flight, called from a business trip in Vienna to voice objections. The team, instead of shouting Mueller down, promised to work to resolve his concerns.

Consensus

Tom Paine, NASA's deputy administrator, polled everyone in the room individually to make sure nobody felt they were being pushed into a decision. The voices around the table spoke confidently, they had reviewed all of the technical and operational challenges — the mission was ready to go. Paine said he felt reassured and would speak with Mueller himself about what to do. Mueller still had reservations about moving so quickly but said he would authorize the planning as long as NASA didn't make a public announcement committing to Apollo 8 until after Apollo 7 flew successfully.

Low kept building consensus, meeting with the government and contractor workforce to discuss the idea and thoughtfully listen to their concerns. His biography recalls a tricky August 1968 conversation with a senior manager at Rockwell, Bill Bergen, who was concerned that the mission might not be ready in time. Low and Bergen together went over the spacecraft systems, with Low reassuring Bergen that only minor changes would be needed to get Apollo 8 ready for lunar flight. Kraft also went over each operational part of the mission and gave his approval to move forward. With Kraft's buy-in Slayton went to talk with Apollo 8 commander Frank Borman to see if he would be interested in going to the Moon.

Borman said he could do it, as long as NASA promised not to overload the crew with extra tasks. "Maybe half a dozen of us sat in Chris Kraft's office one afternoon and we went over the flight plan, to try to understand what would we do on . . . the whole flight," Borman said in a NASA oral history, saying everyone was able to agree to a basic flight plan and a safe mission length by promoting careful discussion of everyone's point of view. "I've always thought, again, it was an example of NASA's leadership with Kraft and their management style that we were able to hammer out, in one afternoon, the basic tenets of the mission."

Ready for the Unknown

With management on board, and the technical details worked out, it was time for the crew to get ready for the mission. Train like you fly, fly like you train is a common mantra for NASA astronauts. In the two months before launch, the Apollo 8 crew focused hard on the last-minute details.

They spent many hours in simulators going over the trickiest parts of the mission, including the fateful trans-lunar injection that would bring them to the Moon and the trans-Earth injection that would bring them home. They not only went over what went right, they reviewed the many ways it could go wrong. They simulated problems on the far side of the Moon when out of radio contact with mission control, such as oxygen leaks and engine problems. And they did egress training at the launch pad and in the ocean, so that when it came time to get in and out of their spacecraft, they relied on muscle memory for the routine parts — freeing their mind for any contingencies.

The astronauts also prepared themselves and their families for the possibility that they wouldn't return, although Anders calculated the odds were pretty good. "I knew every little wire and relay in that Saturn and in the command module. And, as an engineer, [I] probably was [more] able . . . than a lot of the other guys to determine whether it was safe or not," Anders said, adding, "[This] all made me decide that . . . there was one chance in three that [we] wouldn't make it back, that there was probably two chances in three that we wouldn't go there either because we didn't make it back or [we had to abort], and one chance in three we'd have a successful mission."

Strength, Courage and Confidence

> "You gain strength, courage and confidence by every experience in which you really stop to look fear in the face. You must do the thing you think you cannot do." — ELEANOR ROOSEVELT

It was a powerful moment when Borman, Anders and Lovell lifted off from Florida on December 21, 1968. The occasion was so momentous that Charles Lindbergh, who had flown the Atlantic solo a generation earlier, attended the launch with his wife, Anne. Lindbergh, ever the aerospace engineer, famously inquired of the crew how much fuel they would use in their launch. When told, he announced that in the first two seconds, they would have burned more propulsion than he used during his entire trip from New York to Paris in 1927.

In a testament to the training, the Apollo 8 crew and mission controllers made the whole thing look strangely routine. They launched into orbit, safely separated from their rocket and made the trans-lunar injection without a hitch. They not only flew to the Moon, but sought to engage the American public. That year had been par-ticularly difficult for America and Borman decided that they would read part of the Bible's Genesis passage on Christmas Eve. While the religious reading brought criticism to the agency, underneath the message was a symbol of unity for a nation torn apart by civil pro-tests and the assassination of Robert Kennedy and Martin Luther King; here at the Moon, they said, we astronauts are trying to rep-resent all of the United States. The reading was a powerful call to connecting the space program with the public, and to the families who were supporting the faraway astronauts during one of the most prominent holidays of the year.

Christmas Day saw another key moment of the mission pass without a hitch, as the crew triumphantly rounded the Moon after an engine firing that pointed them back towards Earth. "There is a Santa Claus," Lovell joked from about 240,000 miles away to his colleagues in Houston.

America had triumphed, but it had done so in a way that brought more people in the world together. The Christmas Eve broadcast by the astronauts, as well as the other mission activities, drew atten-tion from people around the globe. The mission made the cover of *Time* magazine and was said to bring hope to for the future after the troubling events of 1968.

NASA had triumphed. But would the lunar module be ready in time for a 1969 landing? Its readiness would be one of the keys in bringing people to the Moon's surface in just a few short months.

Leadership Insights

- Respectful discourse, disagreement and decision making based upon the best available data are the hallmarks of successful teams.
- Create teams of experts, give them the resources they need to master the challenges they face, let them do what they are trained to do and relentlessly reinforce why they are doing it.
- Build a constituency of supporters when implementing bold new ideas.

CHAPTER 6

Should We Land or Should We Abort?

"The Eagle has landed."
— NEIL ARMSTRONG

As the Apollo 11 crew descended towards the Moon, a strange alarm began blaring in the Eagle lunar lander as Neil Armstrong and Buzz Aldrin barreled towards the surface.[1]

"Program alarm," said Armstrong, glancing at his display. "It's a 1202."

"1202," echoed Aldrin.

A few seconds passed. The crew had abort in their minds as they waited for word from mission control in Houston.

"Give us a reading on the 1202 program alarm," Armstrong radioed once again.

It takes 2.5 seconds for a communication to go from the Earth to the Moon, and 2.5 seconds for a return message. Within seven seconds of his last transmission and only half a minute after the alarm first blared, Armstrong had his response.

"Roger," capsule communicator and fellow astronaut Charlie Duke said from Houston. "We've got — we're go on that alarm."

Gene Kranz made sure the team was ready for any alarm, any contingency. Their simulation training had worked. The use of simulators had improved significantly in preparation for the Apollo missions and every phase of the spaceflight was practiced repeatedly with the mission control team working with the astronauts in flight-like simulators.

Jokingly described as a "diabolical and insidious team"² by former Space Shuttle program manager Wayne Hale, the simulation training team is responsible for preparing the astronauts and flight controllers to handle any situation that could arise in space. **Astronauts do not want to go into space without understanding and testing every procedure with and without malfunctions, to ensure they are ready to handle whatever might occur.**

Ground teams are no different, and their training experience proved critical to the success of the Apollo 11 landing. Armstrong had to ignore four distracting 1202 alarms and a related 1201 alarm while descending. There was no time to discuss what was happening; the crew had to accept and trust without hesitation mission control's assurances that they could proceed.

Besides which, other problems were quickly mounting. Armstrong noticed the LM was proceeding to an automated landing inside a crater, which wasn't a great idea since they couldn't see the ground where they would be touching down. He took control and began steering. Hovering above the rocky terrain, running low on fuel, he searched for a safe landing spot.

"Thirty seconds,"³ Aldrin said, while reading out altimeter markings from his side of the cockpit.

There was silence as Eagle settled onto the lunar surface in the Ocean of Tranquility at 4:17:39 p.m. EDT on July 20, 1969. Between the computer alarms, the need to manually overfly the targeted landing site to find a more suitable location and the fact they landed with 20 seconds of fuel remaining, Duke's comment, "You got a bunch of guys about to turn blue. We're breathing again," appropriately described the tension of the final approach and landing.

The story of how that program alarm was solved clearly goes back to well before that July afternoon. Perhaps the best spot to begin is with Apollo 9.

Simulate

Lunar landing was not a guarantee after the incredible achievements of the Apollo 8 mission. The astronauts had gone around the Moon and yes, NASA appeared to be just inches away from a successful landing. It would have been easy to ramp up the pace once more and to send an untested lunar module to the surface of the Moon for Apollo 9. But the mission managers were committed to doing two key tests of the LM before sending it on a risky landing attempt.

Apollo 9 would test out the LM in Earth orbit. The spacecraft would separate from the command module and then do a firing to gently make a docking close to home. LM pilot Rusty Schweickart would then do a simulated emergency spacewalk to move from the outside of the LM to the outside of the command module and then inside, to get ready just in case a normal docking was impossible.

The mission was a success, except for Schweickart needing to reduce his spacewalk activities due to routine space sickness, something that most space travelers face during their first few days in orbit that is addressed today with medication and a reduced schedule. One unsung hero of this mission was the team that created the LM simulator on Earth so that Schweickart and Commander James McDivitt could successfully fly away from the command module and then simulate the return and docking before doing it in orbit.

"Some people felt that Apollo 9 was like a backwater kind of thing, but it was not at all to us,"[4] said Frank Hughes in a 2013 oral history with NASA. He retired as NASA's chief of spaceflight training after notable work with crews in the Apollo and shuttle programs getting them ready for space.

"As we went into [the mission], we had to make sure now that the LM was really going to go," said Hughes, a simulation expert who worked closely with the crew. "People had been working below the radar in the simulator world, [and] suddenly, [we] had to get these two simulators to talk to each other. When you got in the Command Module Simulator and you're looking across to the visual at the other one, you didn't see a real LM. You saw one of these models in the model house, and they'd be moving back and forth with cameras, and vice versa. In the LM, they'd look out the window and they would see a model of the command module." Despite the relatively rudimentary nature of the simulators, the training helped the Apollo 9 crew to learn, sometimes through failure, how to succeed in space.

Landing not Allowed

All seemed to be ready for the next flight in early 1969 — Apollo 10, the ultimate test run. Led by Tom Stafford, the crew would fly all the way to the Moon and make an approach to within 50,000 feet. The lunar module, affectionately named Snoopy, was a heavier model that was not designed to land on the surface. Happily, though, it could do everything but — including making several of the crucial firings to get close to the surface, and then return to its other spacecraft.

The crew trained hard for the opportunity, Hughes recalled. "Apollo 10 was interesting. They were pushing the edge of the real flight software now, and they were using it and pushing it to the [greatest] extent possible," he said. "When they flew — and I was very close to that crew, it was a pretty good deal — we had a great time."

That is, everybody had a great time up until the approach took place in space. Apollo 10's two spacecraft separated on the back side of the Moon, out of range from radio communication from Earth. Then the LM began a burn to start to approach the surface. From an initial altitude of 60 miles, they needed to get down to about 9. There were no problems with the descent. Then Hughes recalled what happened when the crew attempted to return to the command module.

"Now they knew just where the command module was, but they didn't know for sure where the LM was," he explained, because the spacecraft had just done a maneuver to go down to the surface. As they initiated the return, "All their tracking filters now kind of blew up," Hughes said.

It was concerning to the crew when the confused LM began spinning about trying to find the right direction to get back to the command module. Cernan, at the controls of the LM, voiced his surprise in the middle of dealing with the problem: "Son of a bitch!"[5] he cried over an open mic. While some in the public weren't too impressed with an astronaut swearing, the moment helped him and Commander Tom Stafford blow off steam and focus. It turned out that incorrect switch settings were causing the LM's erratic movements.

In an oral history in 1997, Stafford joked his mission suddenly became X-rated with that comment.[6] But the crew was able to quickly recover control of the LM, thanks to those hours of simulation training back in Houston. "The whole damned spacecraft started to tumble and tried to rotate," he recalled.

"And real fast, I just reached over and just blew off the descent stage, because all the thrusters were on the ascent stage," he said. The LM was made up of two engines, a descent engine and an ascent engine, and Stafford made the quick decision to simplify by getting rid of the superfluous descent engine. After all, the crew wasn't going to land.

"We got it squared away in about 20 seconds," he added, and the crew did what all crews are trained to do — quickly recover from mistakes and not linger on them, because the dwelling can kill you. Cernan and Stafford once again lined up with the command module and tried to move in that direction. "We did a perfect rendezvous and got all squared away," Stafford said, and the mission moved on.

The sources differ as to how much danger the crew was in, with some saying the crew was seconds away from losing total control, while others said the crew had quite a bit of lead time before running into a real problem. The real lesson learned, though, was to double-check the procedures in the simulations. A wrong switch configuration had led to Apollo 10's problems; for Apollo 11 to succeed everything would have to be done flawlessly.

The World Watched and History Was Made

Gene Kranz was worried in the weeks leading up to Apollo 11, especially when one of his flight controllers made the wrong call during a simulated landing. Kranz was the flight director for the mission control team that would be bringing the crew down to the Moon. During a simulated landing early that summer, Steve Bales — the guidance officer for the mission — called for an abort when the simulated Eagle computer got overloaded.

It was the wrong call, the team learned in the debrief — that spacecraft could have landed despite the problem. Bales and the backroom team supporting him were told to do better. Among the people supporting Bales was Jack Garman, a young computer engineer.

"Those of us in the back room didn't think anything of it. Again, we weren't in touch with the seriousness of simulation to the real world. [We thought,] 'Okay, well, do it again,'"[7] Garman said in a 2001 oral interview with NASA. "But Gene Kranz, who was the real hero of that whole episode, said, 'No, no, no. I want you all to write down every single possible computer alarm that can possibly go wrong.'"

Resilience, Perseverance and Preparation

You can imagine how the teams must have felt. It's weeks before the first crewed landing and you want us, already working overtime, to go through a set of obscure alarms just in case this scenario happens again? Kranz, fortunately, was a leader who led by example. He also was working long hours. He also took responsibility when things went wrong. He had created an open culture within his team where people admitted mistakes, learned from them and talked about how to move forward. Kranz would always be tough, but fair, in listening to his team.

So Garman and Bales, despite the new demands placed upon them, prepared their list of computer alarms. The men knew the LM computer system well and made sure they knew exactly what was happening as each alarm took place.

"The computers . . . are running in these cycles, two-second cycles, for calculating how to drive the engines," Garman explained. "When it got to, I forget what [exact] distance from the lunar surface, the conclusion had early on been that more precision was needed so the computer programs would double their speed. They'd run once a second. . . . You know, as you're getting closer to the ground, the lunar surface, you don't want to coast for two whole seconds. You want to get a little more precision. Everybody knew the computers would be a little busier as a result of that."

And in that list of computer alarms were the 1201 and 1202, which worked like this, Garman said: "One of the test alarms that was in there was one that said if it was time to start the next cycle of calculations — open your eyes, look, calculate, and so on — if it was time to start the next cycle and you were still in the prior cycle, there's something wrong. This is like when you have too many things to do. It's called bow-waving all those tasks; you're not going to get them done. That's not good. So the computer would restart. That makes perfect sense. Flush everything, clean it out, look at those restart tables, and go back to the last known position and proceed forward."

Bales and Garman had printouts of the alarms ready when the Apollo 11 crew made their descent to the surface. Garman said when the alarm happened, the crew — who had not participated in the computer alarm simulation — got a "master caution and warning" alarm that "was like having a fire alarm go off in a closet," Garman said. "I gather their heart rates went way up and everything," he added. "You know, you're not looking out the window anymore."

Bales asked Garman, who had more experience with the computer codes, what was going on. Quickly, the two men glanced down their lists and reminded themselves of the 1202. "If it doesn't reoccur too often, we're fine, because it's doing the restarts and flushing," Garman said in the oral history, and gave the okay to keep going.

"Bales is looking at the rest of the data," Garman said, recounting the story. "The vehicle [Eagle] is not turning over. You couldn't see anything else going wrong. The computer's recovering just fine. Instead of calculating once a second, every once in a while it's calculating every second and a half, because it flushes and has to do it again. So it's a little slower, but no problem. It's working fine. So, it's not reoccurring too often. Everything's stable."

A few minutes later, a 1201 alarm came up and in the enthusiasm of the moment, Garman recalls yelling into the sensitive mic to make sure Bales heard him. "Same type!" Garman cried, and he said he could hear his voice echoing in the loop before Bales himself repeated, "Same type!" A moment later, Duke relayed the message to the crew, repeating word for word: "Same type!"

Garman laughed in recalling the story. "Boom, boom, boom, going up," he reflected on his message from a small back room in Houston, proceeding up the command chain all the way out to the Moon. "It was pretty funny."

Some Things Can't Be Simulated

Sitting in the back room, Garman was just as transfixed as the others while watching the men safely alight on the Moon. It had been simulated so many times that, besides the alarms, Garman said that the whole thing felt strangely normal. But then came the call, moments before landing, that reminded everybody that this was actually the real thing.

"We'd watched hundreds of landings in simulation, and they're very real," Garman said. "On this particular one, the real one, the first one, Buzz Aldrin called out, 'We've got dust now.' We'd never heard that before. You know, it's one of those, 'Oh, this is the real thing, isn't it?'"

Reflecting on the moment, Garman said that solving problems during a descent was something that often happened in simulations. The dust comment, however, set this descent apart from all of the groundwork the team had done in the months before the landing.

"And you can't do anything, of course," Garman said. With Armstrong steering the landing in the last few feet, all anyone could do in mission control was watch, he reflected, "You're just sitting down there. You're a spectator now. Awesome. Awesome."

While the world was transfixed by the astronauts landing on the Moon's surface, Garman said the work to understand why the alarms occurred didn't stop in the back rooms. Anyone who knew anything about the computers, he said, was "trying to figure out what the hell happened."

It was the hardware group in instrumentation who found the problem — the rendezvous radar was in the wrong setting and had overloaded the computer. It was a busy few hours as the Apollo 11 mission teams changed their approach to ensure the same thing wouldn't happen during ascent. Luckily, no other landing crew ever faced down those infamous 1201 and 1202 alarms. Upon retiring, Garman received a T-shirt gift printed with those numbers, in honor of his contribution.

"You don't realize until years later, actually, how doing the wrong thing at the right time could have changed history," he reflected in the oral history. "I mean, if Steve had called an abort, they might well have aborted. It's questionable. That is, those guys were so dedicated to landing that they might have disobeyed orders if they didn't see something wrong.

"But nonetheless," he added, "paths not taken. You have no idea what might have happened. That was . . . in retrospect, one of those points where you were . . . a witness in the middle of something that could have really changed how things went. So it was very good that there were people like Gene Kranz and Steve Bales and others that kept their heads on and thought about it."

"I didn't feel like a giant. I felt very, very small."
— NEIL ARMSTRONG

In his humility, Garman underplayed his role but he thought about it, too, letting the Apollo 11 crew make history in front of the world. Just hours after their dramatic descent, Armstrong and Aldrin stepped out of their hatch and made their famous Moonwalk, spending a little over two hours hopping on the surface in their spacesuits. With their mission accomplished, the crew successfully blasted off the following day and made a landing back on Earth on July 24.

It was a moment of history that billions had witnessed, and thousands had made possible within the walls of NASA centers and at contractors and institutions around the world. Among the leadership moments at NASA, landing humans on the moon and returning them safely to Earth will resonate as perhaps the single greatest achievement in the agency's history.

For many within NASA it was a new beginning, a time to explore farther into space. There were more missions planned that would make even greater demands on the crew, to learn more about our closest neighbor. It was a time of excitement and few would have predicted that the nation would lose interest so quickly.

Leadership Insights

- The value of simulation training extends beyond the aerospace sector. It is useful in a number of corporate scenarios from tabletop discussions of the outcomes of management decisions to testing new technologies and training to operate complex devices.
- Simulation creates an environment where failure is a safe opportunity for learning.
- Develop procedures around a table and test them in the simulator. If they work, train them relentlessly and follow them.

CHAPTER 7

They're in Space Again?

"The flight was extremely normal . . .
for the first thirty-six seconds."
— PETE CONRAD

By the fall of 1969, there were indications that the American public was losing interest in going to the Moon. The war in Vietnam and social unrest continued to dominate the evening news. "Within the United States there was less interest than there was during the summer for Apollo 11,"[1] commented Teasel Muir-Harmony, Apollo curator of the Smithsonian Air and Space Museum, in an interview with BBC. "But around the world there was still quite a lot of enthusiasm for Apollo 12." NASA hoped that providing live color TV signal from the lunar surface would rekindle media and public interest in the program.

In contrast, Apollo 11 was an achievement for the ages. The team overcame serious computer issues to get the astronauts on the surface. The landing was watched by billions around the world, and the entire Apollo 11 crew were celebrated with a world tour and a ticker-tape parade in New York City after their safe return. It was a hard performance to follow.

NASA still had plans to send several more crews to the Moon, to deepen scientists' knowledge of the lunar surface and to eventually practice for more ambitious long-term missions in Earth orbit and, perhaps, farther out in the solar system. Getting the other landings done would require a sustained commitment to excellence and teamwork.

Listen to Experts

The teams supporting Apollo 12 had managed many complex problems to make the mission a reality. Unfortunately, the mission did not get off to an auspicious start when the launch took place into an overcast sky with storms having recently passed through the area. Shortly after launch the rocket was hit by two lightning strikes. "We generated our own lightning,"[2] said flight director Gerry Griffin. "Ionisation of the extremely hot exhaust from the Saturn 5 created a ground." In effect, the rocket became a giant conductive rod connecting the electrically charged clouds with the Earth below.

Commander Pete Conrad reported to mission control, "Okay, we just lost the platform. . . . We had everything in the world drop out."[3] It was largely due to the quick thinking of a mission controller, John Aaron, that Apollo 12 flew safely into space. Aaron realized how they might be able to regain accurate sensor data by switching to the auxiliary data source.

Months earlier, as Aaron recalled in a 2000 oral interview with NASA, "I happened to be on the third shift one night watching a test of the command module that they were performing at Kennedy [Space Center]. Because the [still learning] operators on the third shift at Kennedy were . . . not the 'A Team,' they had gotten themselves in a sequence where they dropped power on the vehicle."[4]

The training Aaron had on the electrical system of the command module made him think that the command module computer would reset all the displayed numbers to zero after rebooting from a power loss. But that's not what was popping up in the various readouts. He recalled seeing numbers such as 6.7, 12.3, "some squirrelly kind of numbers."

After helping the technicians reset the computer, Aaron drove home asking himself where the numbers came from. The following morning, he set a meeting with a North American Aviation engineer called Dick Brown to go through the command module's miles of circuitry, to make better predictions about when that particular pattern of numbers would come up.

Brown and Aaron found this number pattern tended to originate from the Signal Conditioning Equipment (SCE) system, which converts signals from the spacecraft sensors into voltages that would be used for the spacecraft displays and telemetry encoders. SCE was an obscure system, but Aaron followed his curiosity — which he said is a good trait of any person in the space program — to learn a little more about how it worked.

Prepare for the Unexpected

The way to reset the system if it suddenly lost power, he found, was to switch it to an auxiliary setting, allowing the SCE to work in low-voltage conditions.

"[I was] never thinking that when lightning struck the vehicle on Apollo 12, that exact pattern showed up. So it wasn't that I understood exactly what had happened, I recognized a pattern and how to get out of it," Aaron said.

Aaron, looking at the Apollo 12 readouts, then said a command to his mission control team that sounded more like a magic spell: "Flight [director], try SCE to Aux." Flight director Gerry Griffin had never heard this obscure sequence, and neither had the capsule communicator, astronaut Gerald Carr. Carr asked Aaron to repeat what he had said, but remained confused: "What the hell's that?"

But with the rocket accelerating along its trajectory, Carr had to trust Aaron and he relayed the call to the spacecraft. Of the three men, the lunar module pilot Alan Bean knew where the switch was and Bean continued the chain of trust that had started with Aaron's call to Griffin and Carr. He quickly flipped the appropriate switch in the command module and the capsule came back online allowing Apollo 12 to make a safe entry into space.

Trust that had taken months of hard work to build was already at play in the early moments the Apollo 12's flight and mission control double checked the data to make sure they were confident in the spacecraft's abilities to move forward. Everyone carefully reviewed the data displayed on their monitors from Apollo 12's systems again before giving the okay to do the trans-lunar injection to head out to the Moon. This incident reinforced the awareness that mission controllers really had to know the insides of their spacecraft — an awareness that, as it turned out, would help save Apollo 13.

But that was in the future. For now, Apollo 12 was on its way to the Moon and, unlike Apollo 11, the lunar descent got off to a great start with no computer errors and a pinpoint landing. Excited, the crew made their way onto the surface equipped with the color video camera to overcome the shadowy images TV viewers saw on Apollo 11.

Slow Down to Speed Up

Conrad, focused on getting started with the lunar activities, urged his crewmate and close friend Bean to "Hustle, boy, hustle," because "We got a lot of work to do."[5] Even though the men were close friends from before their NASA days, Conrad was very much the senior commander on the mission. He had flown in space before and had recommended Bean's appointment to the mission. NASA management had stuck the rookie astronaut in an obscure department known as Apollo Applications, and in numerous interviews the humble Bean has credited Conrad with getting him a ticket to the Moon's surface.

Conrad was a highly experienced astronaut who, among other milestones later in his career, helped save the Skylab space station from overheating. But in telling Bean to "hustle," he had inadvertently increased the risk of error. Telling Bean to move faster on the surface did work, but it also may have contributed to a crucial mistake.

Bean reduced the planned five-minute familiarization period to get the camera set up. He was feeling reasonably comfortable on the lunar surface and felt confident pushing the timeline, not worrying about the camera setup because it was within his control and he had practiced it on Earth.

Bean's checklist had a warning to keep the camera away from the sun during the initial TV panorama: "omit up-sun." Yet the feed from the camera, according to the Lunar Surface Journal, shows shadows suggesting the camera was indeed looking up-sun. Moments later, the sun came directly into the field of view.

Mission control warned Bean as fast as the team could: "Al, we have a pretty bright image on the TV,"[6] said capsule communicator Ed Gibson. Bean, heeding the call and working outside of the training sequence, may have accidentally pointed the camera at the sun a second time, the Journal adds. The astronauts and the ground team spent quite a while trying to fix the camera before deciding it was best to give up. The mission continued, but the lack of television caused TV viewers and networks to carry less coverage of the mission — likely contributing to an apathy seen in Apollo 13's initial coverage.

"I never really thoroughly understood the limitations of the TV,"[7] Bean said in a post-flight debrief on the incident. He took full responsibility for his actions while making constructive suggestions that let all other crews have a working camera on their missions.

"I think that the way we can help a situation such as that, in addition to doing a lot more pre-flight thinking about it, is to get a TV to work with [during training] that's like our flight TV," Bean continued. "We need to work with it outside in the sun using the monitor. If we had done this, I think it would have become very obvious that the TV doesn't have to be in the sun too long — or even point[ed] at a bright object too long — before the tube is going to saturate and you're going to run into a lot of trouble."

Debrief

From the early days of the Mercury and Gemini programs, post-flight debriefs were a critical opportunity to prepare for upcoming missions. The lessons of what worked, and what didn't, helped the astronauts and mission control teams decide if changes needed to be made.

The first recommendation was to use the post-flight quarantine period to get as much done as possible, allowing the astronauts to take a well-deserved rest with their families after emerging from quarantine. The Apollo 11 crew were kept busy by the media during quarantine and were whisked off to worldwide celebrations immediately after it finished. This was great for the program but put a strain on the astronauts and their families. Aldrin, for example, fell into a depression and alcoholism, and later said he wished there had been some time after the mission to decompress. With Apollo 12, there was considerably less media attention and the crew used the time for technical debriefs and reports, emerging from their isolation relaxed.

"It gave us the opportunity to write all of the reports that were required of us — all of the pilot reports. In addition to that, all the briefings,"[8] Gordon said in a 1997 oral history with NASA. "Between a biological barrier, we briefed other flight crews, we briefed mission control, Flight Control Division, system engineers. So once we got out of quarantine, we were all done. We didn't have anything more to do. Apollo 12 was over with except for the studying of the lunar materials that [were] brought back."

"It was an appreciated time," he added. "Even though we felt that we didn't need to be in quarantine, we used it to our advantage. And by the time we got out, all the reports were written, all the debriefing had been accomplished and that was it."

The importance of Apollo 12's precision landing cannot be overstated. The crew made it down a little off target because the immediate landing zone was unsuitable, but close enough that the Surveyor 3 uncrewed spacecraft — one of their destinations — remained within an easy walking distance. To achieve the scientific objectives of the program NASA was planning to target highland landing sites starting with Apollo 13, rather than the huge plains that Apollo 11 and Apollo 12 landed upon.

Understanding the Past

Getting to those highlands would provide crucial information about the Moon's early history, but it would also require precision landings with less room for error. Apollo 11 and 12's crews came back to Earth with mostly the same type of basalt and igneous rock samples from the same time period in the Moon's history. The scientists, however, had bigger questions they wanted to answer about how the Moon formed that would require a larger variety of samples from earlier in its history. The highland missions of Apollos 14, 15, 16 and 17 succeeded, bringing back older Moon rocks, including one dubbed the Genesis Rock — a 4.1-billion-year-old anorthosite igneous rock showing what rock types were present early in the Moon's history.

From the work of the various highland missions, scientists theorized that the Moon formed after a Mars-sized world crashed into the Earth billions of years ago. The resulting debris created a vast dust ring circling our planet, which coalesced gradually into the Moon we see today. Some parts of the theory are still confusing scientists today, such as why the Moon is so metal-rich in the lowlands and so metal-poor in the highlands compared to Earth, but new theories are emerging thanks to the work of the Lunar Reconnaissance Orbiter and other missions.

Leadership and Teamwork

Another key lesson from the mission was the importance of teamwork and the benefit of Apollo 12's high crew cohesion that arose from the longtime relationships between the astronauts. Bean first met Conrad at the U.S. Naval Test Pilot School in Maryland, where Conrad was his instructor. Conrad and Richard Gordon were past roommates on the aircraft carrier USS Ranger and also worked together on a past crew, Gemini 11.

> "As a general rule, on every flight you go somewhere that you get to share your story and your experiences with people who helped you get there." — CHARLIE BOLDEN

During the space station era, NASA implemented leadership and team training that enabled crews to spend time living and working together before going into space. The expeditionary behavior training of the shuttle crews and long duration astronauts included a program from the National Outdoor Leadership School that gave astronauts an opportunity to use wilderness expeditions to develop their team skills on Earth, before going into space.

NASA also added more analog environments to help familiarize unflown astronauts with the rigors of working in extreme environments. The NASA Extreme Environment Mission Operations (NEEMO) missions that started in 2001 continue to send astronauts

60 feet underwater to the Aquarius undersea habitat, just off the coast of Key Largo, Florida. Astronauts not only develop their skills working in teams, but practice using technology and techniques for missions on the International Space Station and future lunar missions.

For space station crews, NASA joined with the European Space Agency in sending astronauts on CAVES (Cooperative Adventure for Valuing and Exercising human behavior and performance Skills) missions. This program sends astronauts into caves in Sardinia, Italy, to perform science experiments in a dark environment where it is tough to distinguish day from night.

NASA also runs other analog missions to learn more about how people behave in isolated environments. They have used research from habitats in Hawaii, Utah, Canada's Devon Island, Russia's Institute of Biomedical Problems and NASA Johnson Space Center's Human Exploration Research Analog (HERA), to further understand and better predict how to keep crews safe and healthy while working together for weeks or months at a time in space.

Perhaps Apollo 12 would have been more memorable had they not recovered from the data loss following the lightning strike. In the competitive arena of nightly news coverage, the success of the mission was lost among the other competing stories. The crew came back to Earth from a mission where just about everything went right, where they quickly recovered from what could have been an aborted mission following the lightning strike. The only problem was the troublesome failure of the lunar surface camera, hardly enough to garner national media attention. The relative ease of the highly trained crew and mission control team in managing the challenges of Apollo 12 might have made the nation more complacent. Perhaps bad news really does get more attention than good news. However, Apollo 13 would soon test NASA to a degree it never had experienced before and recapture public interest.

Leadership Insights

- Speeding up can lead to errors that take time to resolve ultimately slowing operations down. Slow down and focus to speed up.
- Time is an important asset: use it efficiently to optimize success.
- Leadership training, team building and developing the behavioral competencies of individuals and teams is an important adjunct to technical training.

CHAPTER 8

Failure Is Not an Option

"We in effect framed up the flight return
that we were going to have."

— GLYNN LUNNEY

Space can never be a routine matter. The environment is inherently dangerous, the technology cutting-edge. Mistakes can be costly in terms of money and, at worst, in terms of life. But perhaps we all forgot that as the third Moon mission launched on April 11, 1970.

Apollo 13 certainly felt routine at the beginning. Fewer television stations and newspapers covered it, and for those who did there was much less attention to the mission. Fewer people tuned in to the coverage. It didn't help that Apollo 12's camera had failed so early in the previous mission, and people hadn't had a chance to experience the perspective of visiting another location on the Moon.

The sentiment now was approaching "been there, done that." Initially the Moon-landing program was supposed to go all the way to Apollo 20, but funding for that was cut in January 1970. Apollos 18 and 19 were in danger of disappearing forever; but nobody knew that for sure when Jim Lovell, Fred Haise and Jack Swigert launched to the Moon at 13:13 local time on Apollo 13. The cancellation of those missions was confirmed later in the year.

Last Minute Changes

It had been a troubled last few weeks for the crew. Ken Mattingly was the command module pilot assigned to the mission, but Mattingly was unwittingly exposed to the German measles shortly before the flight when fellow astronaut Charlie Duke's son fell ill. Mattingly had no natural immunity, the doctors said. The crew would need to launch with the backup, Jack Swigert.

Lovell did his best to make Swigert feel accepted. The crew only had two days to train as a group before lifting off, and Swigert had been out of the loop recently as a backup crew member. The backup astronauts were typically more occupied with finding hotel rooms for visiting family members than training before launch, Lovell recalled in a 1999 oral interview with NASA.[1] But Lovell's fears were addressed as soon as Swigert sat down with Lovell and Haise.

"Jack happened to have written the malfunction procedures for the command module. So, he knew the command module pretty good," Lovell recalled. One challenge was quickly building team cohesion, at a moment when Haise and Lovell could read the inflections of Mattingly's voice during critical procedures. But Swigert was well trained and ready to go.

"The two days we worked with Jack, he appeared perfectly comfortable with the vehicle," Lovell recalled. "And so, I said, 'Go.' Because they [senior NASA management] came to me privately and said, 'Are you happy? Are you satisfied? Do you want to go?' I said, 'Sure.'"

There were a few technical blips on the way to the Moon, as things always go, but there were no showstoppers and the new crew appeared to be doing just fine as they completed the trans-lunar injection and began their journey to the Fra Mauro highlands. On the evening of April 13, the crew performed a public relations event — a TV broadcast that unfortunately few got to see live. But what happened next quickly got beamed around the world. Lovell recalled, "I think it was either nine or ten o'clock back here in Houston. And I'm coming back down through the tunnel, and suddenly there's a hiss-bang! And the spacecraft rocks back and forth."

Cascading Events

The liquid oxygen tanks on the Apollo command module were designed to be stirred by an electric fan inside the tank to get accurate readings on the gauging systems as the cryogenic oxygen tended to stratify in the tanks.[2] NASA had developed a procedure to stir the tanks and accurately track the quantity of oxygen for the crew to breathe and also for the fuel cells that ran on hydrogen and oxygen.

The procedure was called a "cryo-stir," and it was fairly simple as routines go. A crew member — usually the command module pilot — would flip a switch in the spacecraft to turn on the fans. The fans would gently agitate the cryogenic (super chilled) oxygen and thus remove the layers. In 13's case, this procedure had happened four times before, and it was one of the last things planned before the crew's bedtime.

This time, something different happened in tank No. 2. It exploded, not only letting go the crucial oxygen, but also tearing apart vital connections needed to keep the command module functioning. The master alarm on the spacecraft blared and the crew made their famous call for help: "Houston, we've had a problem."

Even in this moment of emergency, the crew went through procedures, including working to diagnose the problem and to "button up" (or close) the hatch to the attached lunar module. At this point they didn't know what had happened and were concerned that a meteor had hit the spacecraft. The astronauts and mission control worked quickly to understand what had caused the master alarm. "We didn't know, 'Is that an . . . instrument problem?' Because obviously we [didn't] lose all the oxygen," Lovell recalled. After all, the crew was still breathing fine without spacesuits. "And, you know, this went back and forth," he said. The initial thought focused on a sensor error, but as more alarms sounded it was clear it was something critical.

The troublesome tank had first been installed on Apollo 10, then removed for some modifications. The tank extraction crew accidentally dropped the tank two inches, jarring an internal fill line. At the time, nobody knew the line had been damaged. The tank was pulled off Apollo 10 for an external inspection and reinstalled on Apollo 13.

Of course, there was more testing on the ground for Apollo 13. The tank wouldn't empty correctly, which in retrospect was probably due to the damaged line. Technicians did their best to get around the issue. Their solution was to "boil off" the oxygen instead of draining it by using the tank heater. Although the procedure was closely monitored, there were other design issues that turned No. 2 into an accident waiting to happen.

Something Amiss

The first had to do with voltage. The oxygen tanks were designed to run off 28-volt DC power in the command and service modules, but a redesign also let them accept 65-volt DC ground power at the Kennedy Space Center. Everything was upgraded — everything except for the heater thermostatic switches, which were overlooked.

The heaters in the tanks were normally used for very short periods to heat the interior slightly and increase the internal pressure to keep the oxygen flowing. The decision to use the heater to "boil off" the excess oxygen in tank 2 required eight hours of 65-volt DC power. The post-flight investigation concluded that the prolonged heating probably damaged the thermostatically controlled switches on the heater that were designed for 28 volts causing them to weld shut. This may have allowed the temperature within the tank to rise to over 538 degrees C (1000 degrees F). The high temperature wasn't noticed as the sensors inside the tank were designed to measure to only 27 degrees C (80 degrees F). Baked in this extreme heat, the Teflon insulation surrounding the electrical wires eroded away. What was left behind were wires in pure oxygen — a dangerous situation just waiting for one stray spark inside the tank.

All this wasn't known to the Apollo 13 crew, or to mission control, when the master alarm blared in their spacecraft. Within minutes of the incident, however, everybody knew that it couldn't be instrumentation.

"Something was wrong with the electrical system," Lovell recalled of the explosion's aftermath. "We eventually lost two fuel cells. We couldn't get them back. Then we saw our oxygen being depleted. One tank was completely gone. The other tank had started to go down. Then I looked out the window, and we saw gas escaping from the rear end of my spacecraft."

Apollo 13 transformed from a Moon-landing mission into a dire emergency. There had been previous emergencies in space but nothing to this degree. The team in mission control would have to work closely with the crew to find a solution. The crew had already lost the chance to walk on the lunar surface. The next real danger was losing the crew, leaving Swigert's parents without a son and Lovell's and Haise's families without husbands and fathers. Haise's wife Mary was pregnant again, and Haise — in his moments of reflection — wasn't sure he would ever meet their new child.

Back in Houston, Flight Director Gene Kranz's White team was on console, communicating with the astronauts in space. He fought to keep his team focused on solving problems as quickly as they came up. "What you got to do is figure out what is the problem,"[3] he recalled in a 1999 oral interview with NASA. "We spent almost 15 minutes before we finally concluded we had an oxygen tank go south on us. If we'd had today's technology, we'd have picked that up literally in seconds. **So what is the problem? Then you can figure out what are you going to do about it, which direction you're going to go."**

Developing a Return Plan

Just like the astronauts, the mission control team goes through extensive training before any mission launches. Kranz was lucky to have a team of veteran controllers with him, as young as they were. As they began piecing together how to save the astronauts, they thought about what procedures they might have. An idea flashed in the team's memory — something they had trained for in late 1968 and early 1969 in the run-up to the Apollo 9 mission, which tested the lunar module.

> "This crew is coming home . . . and we must make it happen."
> — GENE KRANZ

"Part of the training, one day my team didn't do the job right," Kranz recalled in a 1999 oral interview with NASA.[4] "When we were debriefing the training, our Sim Sup [simulation supervisor], which is our training boss . . . comes to us in the debriefing and says, 'Why did you leave the lunar module powered up? Why are we using all that electrical power? Don't you think you should have developed some checklist to power this thing down? Whenever you've got trouble, you ought to find some way to conserve every bit of energy, every bit of resources you've got, because some day you might need it.'" This was the beginning of what Kranz would later call "the lifeboat procedure" to evacuate the command module and temporarily use the lunar module as the lifeboat.

The team thus had an attack plan for saving Apollo 13. They hastily sent up procedures to the crew to shut down the crippled command module and to power up the attached lunar module, luckily left undamaged in the explosion. Kranz, however, knew his team would be in this for the long haul; it would be several days before they could get the astronauts back home, given they were approximately 200,000 miles (322,000 km) from Earth and still heading towards the Moon.

Fortunately, NASA had already set up a rotating set of teams and had handover periods between teams where each set of controllers at a workstation would debrief and get ready for the next shift. This worked beautifully even when things were going nominally, Kranz said. "Basically, we were able to distribute this very large base of knowledge and get it packaged in four flight directors," Kranz said, providing crucial checks during a mission at vital decision points.

It was decided, among Kranz and other mission managers, to pull most of his team off the consoles and to form what was known as a "tiger team" to attack the "consumables problem." They had to figure out how to stretch the lunar module's oxygen, water and carbon dioxide scrubbers for a four-day flight home with three people instead of a two-day lunar mission with two people.

A quicker direct abort, NASA decided out of caution, would not happen because there was no guarantee the command module engine would work after the explosion. Instead, the astronauts would use the less powerful LM to swing around the Moon, borrow the Moon's gravity and boost the spacecraft stack back towards Earth as swiftly as safely possible.

Every Idea Is Welcome

It was funny, Kranz said. He encouraged the team to check past training records and to pull from their memories in coming up with the right solutions. Even though the team had never faced such a problem before, the training and procedures developed through Apollos 7 through 12 still proved useful.

"There wasn't anything, no matter how obscure, no matter how way out, that we didn't look at and say, 'Hey, we might be able to use this downstream. So, let's take it, write it out completely,' and we'd . . . basically establish the center procedures for this case, and then we'd put it sort of like in a bookshelf in a library. In desperation, when time's short, you want to go back to something that you've known and maybe tested before as opposed to try inventing on the spot. And our lifeboat procedures were part of that package."

Glynn Lunney, flight director of the Black team, had his group of flight controllers work on the consumables problem and a plan to return home. During a 2020 Space Center Houston event, he recalled how his team calculated the load on the command module's electrical system, particularly as the crew got closer to the Moon.[5]

"You can be very, very conservative about that, or you can be more realistic — and the more realistic answer was that we could power the [command module] vehicle. We could power it up and take a burn, when we get around the corner of the Moon, and go around the corner . . . on the way back home, then we could power it down. That would leave us quite a while to save on the power, however, as a measure of how conservative we were."

Team cohesion was crucial as the controllers kept a watch on the spacecraft's consumables problem. In a June 2020 interview, Lunney said he "didn't do anything special" to keep the controllers on track but pointed to their long history of working together; some had been there since Gemini, and of course Lunney himself had been flight director before Apollo 13. "We just didn't have any guys in there that shouldn't be there. And we did pretty well with it," he said.

Mattingly, who became a CAPCOM for the mission after the measles scare, commented that Lunney's leadership was the finest he had seen. "Glynn walked in there, and he just kind of took charge," Mattingly said. "At that point, nobody would even think of saying anything about disasters. . . . it's just professionalism at its finest.

That was all exclusively caused by one Glynn Lunney. Absolutely the most magnificent performance I've ever watched."[6]

The emergency had taken place towards the end of Kranz's shift and fortunately Lunney had come in ahead of schedule to get ready for what he thought would be a routine handover. "I worked for about six extra hours on the front end of my shift and it was quite a time," Lunney said, and his team kept focused on solving problems as they arose, just as Kranz's team did.

"I just was so very proud of the work that we had done. And we got the vehicle settled down. We got it started on its way back home. We got a plan in place," he recalled.

"We had various ideas about what we should do, but we had eventually settled on the plan that we wanted. And I was really comfortable with all the guys that we had, and the work that they were doing and the fact that they were pretty well-trained."

Doing What It Takes

It is worth noting that most of the team were extraordinarily young. Mission controllers of that era tended to be in their late 20s or early 30s. The stamina of youth came in handy during the long hours everyone pulled to bring Apollo 13 home safely. Despite their relative youth, there was enough seasoned experience in the group through past missions to figure out the plan forward by the end of his first shift after the incident, Lunney said. The 12-hour shift was longer than usual, but the team still accomplished a tremendous amount in that short time.

"We in effect framed up the flight return that we were going to have," Lunney recalled. "We had a [consumables] margin on it by the time we ended up with the plan that involved going around the Moon." That plan required an engine burn at the back side of the Moon, and another burn two hours after rounding the back side, to put the spacecraft on track to reach Earth.

There were problems here and there as the crew made the voyage home. Their carbon dioxide scrubbers in the lunar module became saturated and mission control had to help them adapt differently sized ones for the command module. A battery burst. The trajectory drifted off course, requiring another burn. But overall, Lunney recalled, there was a sense of calm in mission control from relying on the procedures and from recalling the way Low had led the recovery after the loss of Apollo 1 just three years before.

"It wasn't like I was had made a big, news-breaking or brand-new thing that had never been thought of before," Lunney said. "We'd done a lot of thinking and planning for various kinds of situations. And it was natural to fall into a position where I knew what we were going to do. We took each step kind of deliberately. And George Low was not the manager at the time; he was in . . . Washington. George was there for a point . . . I think George was very pleased with how it went. He had every right to be pleased because he was there when we were of course recovering from the fire . . . he was comfortable with what needed to be done."

One of the biggest problems they faced was keeping the crew comfortable and engaged in the decision making. At times, the astronauts felt communications lagged — particularly when it came to the re-entry procedures. What helped earn their trust was putting a well-regarded colleague in charge of re-entry. Mattingly, who never got the German measles, stepped in as the CAPCOM and was part of the team getting the command module ready to return home. Mattingly himself transmitted the procedures up to the crew as they prepared for splashdown.

The crew made it back safely, bobbing as usual in the Pacific Ocean. As with any mishap, NASA had to stand down to re-examine its procedures and the spacecraft, before giving the all-clear to launch Apollo 14 in January 1971. The oxygen tank explosion was a tremendous challenge and triumph for the agency that helped everyone appreciate the cascading factors that can contribute to an emergency situation. The performance of the astronauts and mission control got everybody home safely and would result in references to "successful failure," but for the nation it was one of NASA's finest moments.

Leadership Insights

- All ideas are welcomed when seeking solutions to unforeseen situations. Actively managing ideas, implementing the best while keeping others on hold was the key to the success of the Apollo 13 mission.
- It is important to relentlessly build competency, experience and teamwork to respond effectively to emergencies.
- What you believe is what you achieve: identify a goal and work to make it reality.

CHAPTER 9

Been There, Done That, What's Next?

"We need more handshakes in space."
— BUZZ ALDRIN

Apollo 13 had narrowly avoided tragedy, but it was also a stark reminder of the risks associated with human spaceflight. The dramatic rescue of the crew had captured the interest of the nation, but attention soon turned back to the many problems in the news. Budgets were getting tighter and Apollo 20 was canceled before Apollo 13 flew with Apollo 18 and 19 sharing a similar fate shortly after the heroic return of the crew. Four more crews would venture out to the Moon between 1970 and 1972, providing a leadership and scientific legacy that continues to serve as a source of inspiration today.

In the background of the lunar voyages, the political environment was changing. The rhetoric between the United States and the Soviet Union had cooled down. There were no discussions about missiles in Cuba, and negotiations had begun about taking down the nuclear storeholds that had made the Cold War such a frightening time. And what better idea, some people thought, than to show the two superpowers collaborating instead of competing in the exploration of space?

The Soviets never successfully launched their big Moon rocket and had shifted their focus to the more achievable goal of building a space station in Earth's orbit. That ended up being an advantage for the United States as it provided a more cost-effective post-Apollo vision for the program. Some of the Soviet space stations were admittedly military programs disguised with civilian intent. However, as the decade progressed the Soviets were quickly racking up time in space.

The first space station was Salyut 1, designed for a crew of three people. The station spent 175 days in orbit in 1971 but was occupied for only 23 of those days. The Soviets tried sending up other space stations, but three successive attempts failed to reach orbit, or didn't survive long enough to have crews. But starting with Salyut 3 in 1974, the Soviets had a number of successes in long-duration spaceflight.

Time to Pivot

The Americans had a single space station in that era, Skylab, which launched in 1973 and hosted three crews who spent 171 days, 13 hours in space over the course of the three missions. The first crew on Skylab 2 had to do an emergency repair to stop the station from overheating but it functioned flawlessly for the remainder of its 2249 days in space. The longest Skylab crew stayed in space for 84 days studying how humans adapted to living and working in microgravity. In comparison, the Soviet Salyut 3 in 1974 only had a 24-day mission, but within a few years other Salyut missions quickly beat the American mark. The longest flight, Salyut 6 in 1980, lasted 185 days with several more flights that approached or went past the 100-day mark.

As the Apollo program began to wind down, then administrator Thomas Paine wanted to focus NASA's limited dollars on getting ready for a space station. He was eager to embrace the spirit of *détente* and the possibility of collaboration in space. He had already learned a bit of Russian in school after returning from the Second World War, being convinced that learning the Russian language along with understanding nuclear energy were the best things to focus on to prepare for the post-war world. As space spending dropped in the United States, Paine declared that exploring this new frontier was such a huge endeavor that spacefaring nations should work together.

Learn to Collaborate and Collaborate to Learn

"I decided — and I hope I made the right decision — that although [NASA Administrator] Jim Webb certainly had done a tremendous job of building up NASA and the program on the basis of the Russian threat . . . times had changed. The time had come for NASA to stop waving the Russian flag and to begin to justify our programs on a more fundamental basis than competition with the Soviets,"[1] he said in a 1970 interview with NASA. His comments echoed those of NASA's first administrator Glennan who felt that to be sustainable, the program needed to pursue its own vision and goals and not be driven by competition alone.

Paine was not the first to think through a Soviet partnership; even U.S. president John F. Kennedy's administration was pondering the idea way back in 1963, in the months before he was assassinated. The thought of helping each other out, however, was difficult to achieve when the nations were still locked in a race to the Moon. The race was over now. And Paine, despite many failed efforts by past NASA officials to create joint scientific collaborations with the Soviets, wrote once again to the Soviet Academy of Sciences in early 1969 to see about the former rival's interest.

The Soviets were gracious in this early correspondence with Paine, even sending public and warm congratulatory messages after the Apollo 11 crew landed that July. NASA's Space Task Group was working on a report that same year that said, in part, that the Apollo Moon-landing goal created a "feeling of 'oneness' among men [humanity] everywhere . . . that can provide the basis of new initiatives for international cooperation."

> "It was the opportunity to bring people from a really foreign culture to the United States and introduce them to our way of life and help them adjust and adapt and getting to know them as true friends and establishing a lifelong bond with them that exists today." — CHARLIE BOLDEN

U.S. president Richard Nixon told Paine to keep pursuing the idea of Soviet collaboration in late 1969 during a historic moment for NASA; the president and Paine were in Air Force One together, on their way to meet the Apollo 11 crew after their successful splashdown from space.

Following through on the idea, the president created an interagency committee that would look at all aspects of creating joint space missions with the Soviet Union. This committee would present their ideas to the White House and, as it turned out, every agency was on board except the Department of Defense.

Apollo-Soyuz Test Project

To summarize the many meetings, negotiations and the flights back and forth from Moscow would be a lengthy task. Suffice it to say that **collaboration is not always easy,** learning to work together takes time and commitment. Trying to find common ground was difficult but in the early 1970s the vision of a shared space mission emerged as one of the ways by which the American and Soviet programs would work together.

A small team led by Dr. Gilruth was put together to meet with their Russian counterparts to explore ideas for the mission. Lunney, who ultimately became program manager of what would become the Apollo-Soyuz Test Project, remembers the challenges well: "We were going to explore this idea that had been raised about making rendezvous and docking aids sort of mutually compatible . . . it's sort of like having the same kind of lifeboats on your ship so that somebody else could pick them up."[2] The initial meetings were difficult: "It was more like it defined a sense of competition or hostility, adversarial hostility, that existed at the time."

There were concerns on both sides and Lunney remembers the "sense in some of the NASA people who were not involved in it, that there was still a residual of distrust that carried over from their experiences one way or another. . . . They were, you know, objective to neutral about it. They were not overly hostile towards the idea, but they weren't overly supportive . . ." As the former adversaries began working together on the technical aspects of a possible mission the trust and respect grew on both sides. The mission was eventually agreed upon in the Agreement Concerning Cooperation in the Exploration and Use of Outer Space for Peaceful Purposes signed in 1972, leading to one of the most unique collaborations of the cold war. Despite the politics, it was about people. The teams of engineers found common ground in their passion to explore space and the aloofness of the initial meetings quickly turned to friendships.

The mission was ambitious — to have an Apollo spacecraft dock with a Soviet spacecraft in Earth orbit. The Soviets were usually reluctant to share much information about their spacecraft; indeed, they preferred to announce successful launches after the fact in the early days of their program.

The Americans had their own concerns as well. Some wondered was it safe to dock with a spacecraft that had fewer redundant components than Apollo? What would it mean to have Soviet cosmonauts visiting American facilities to train with American astronauts? Were there any military risks that might arise from sharing technological information?

Tom Stafford, who commanded the crew, recalled in a 1997 oral history that the Soviets didn't want to name their crew until six months before the mission. "Hey, this isn't going to hack it," NASA said to the Soviets, in Paine's words. "We insisted on a minimum of two years."[3]

NASA realized the magnanimous thing to do would be to continue to set the tone for how the collaboration might proceed. The agency named its astronauts first in February 1973, three months before the Soviets felt comfortable enough to do so. The American crew plunged almost immediately into Russian-language training to help their counterparts feel more comfortable. Perhaps taking the lead worked better than expected. The Russians quickly hired full-time English professors to follow their cosmonauts around, causing Stafford to complain that "they were skunking us." NASA acceded to Stafford's request to have their own professors available to the crew from early in the morning to late at night — even on weekends.

It wasn't always an easy relationship. The KGB followed the Americans around at times, Stafford recalled, while the cosmonauts had their own way of doing things; on their daily drive in a tour bus to the Russian training center at Star City, Stafford recalled his escorts using a bullhorn, "yelling the peasants off the road." That attitude probably didn't endear them with the locals.

> "The chance to do that [ASTP] with the Russians in 1975, middle of the Cold War, as a young man, was quite a thing and we pulled it off." — GLYNN LUNNEY

Despite these challenges, the partnership worked in principle. The crews trained together. They became friends, and with the possibility that a failed docking attempt could turn into a collision, they trained as a team and learned to trust and respect each other. When they launched their missions in 1975, the docking went off without a hitch. The two spacecraft remained together for about a day, with the crews willingly sharing meals and joint experiments for the TV cameras. They had demonstrated the compatibility of rendezvous and docking systems and created the possibility of an international space rescue. The cynical felt it was purely a publicity stunt, but many at NASA knew that this was the beginning of future collaboration in space.

Soviet relations once again cooled amid nuclear talks later in the 1970s, so a follow-up joint mission never went forward. The two agencies, for a time, went their separate ways. NASA began flying the space shuttle and was busy trying to satisfy the new interagency requirements of this program — such as classified Department of Defense missions to launch satellites. The Soviets continued their long-duration research on Salyut 7, last visited in June 1986 by the crew of the Soyuz T-15 spacecraft in a historic spaceflight between two space stations.

Salyut 7 was launched in April 1982 and was first crewed in May of that year with two cosmonauts arriving on a Soyuz T-5 spacecraft. Research continued aboard the station with various cosmonauts visiting the station until there was a medical evacuation of the crew in November 1985. The station was the most advanced of the Salyut series when it was launched but it was due to be replaced by a much larger modular Mir space station, and for a short period of time the Russians would have two space stations in orbit at the same time.

Mir was a milestone space station. It was the first modular space station to be assembled in orbit and the first core module was launched from Baikonur aboard a Proton rocket in February 1986. Rather than sending a crew to Mir and another to Salyut 7, the engineers came up with an idea of using the same Soyuz T-15 spacecraft and crew for both missions. The cosmonauts first docked with Mir in March spending a month and a half activating the station. They then got back in the Soyuz T-15 spacecraft and transferred to Salyut 7 in early May to complete a number of tasks before undocking and shuttling back to the Mir space station at the end of June. From the ASTP international docking mission a decade earlier to commuting between space stations, it had been a remarkable time in which both NASA and Russia were developing the capability for spaceflight in Earth orbit.

Mir quickly became unique in its operational and scientific achievements. Until 2010, it held the record for the longest continuous human presence in space, at 3644 days. The record to this day for the longest single human spaceflight was set aboard Mir by the nearly 438-day flight by Valeri Polyakov between 1994 and 1995.

The official translation of the Mir space station's name is "Peace." Given the political climate of that era the definition could have been considered either ironic or aspirational. Former space shuttle commander Frank Culbertson considered the importance of the name when giving a mid-90s presentation at the 10th Congress of the Association of Space Explorers. He reflected that as he learned more about Russian history, he understood that Mir could mean: "What I think we would call a village . . . where all the local people lived in close or communal proximity to better share the limited resources . . ."4 The concept of working together in space resonates with most astronauts and cosmonauts as it did for Culbertson. The idea of Mir as a village seemed appropriate to him: "We all love space flight, we all love the engineering challenges, we all love completing the assigned operations, and we have shown we can accomplish operations in space unilaterally, bilaterally, multi-nationally, cooperatively, competitively, upside-down, inside-out, by any gender or race, and occasionally with the most incredibly dubious payloads imaginable. And . . . we can show we're having fun while working hard." Perhaps that definition was the most prescient of NASA and Russia working together again.

Mir's cultural milestones were many, including flying a number of foreigners on short visits during crew changeovers that were part of a growing willingness by the Russian program to collaborate in space. Mir, despite the political tensions of that era, heralded the future of space exploration where international space flight participants could work together in a modular space station open to further scientific knowledge.

There were no Americans involved in those early days due to continued poor relations between the Soviet Union and the United States. But there were a few in NASA that remembered the benefit of the ASTP mission who quietly hoped for the opportunity to work together once again. The U.S. interest in collaborative missions began to build when newly elected President Regan announced that he was directing NASA to develop a permanently crewed space station and invite other countries to participate. For the Reagan administration it was, "A leadership issue very much in the context of the Cold War."5 The Soviets were not invited to join the Freedom project with the other international partners, but the Senate issued a call for renewal of U.S.–Soviet space cooperation with the passage of a resolution noting U.S. readiness to work cooperatively in space "where mutually beneficial" with the Soviet space program.

The emergence of Mikhail Gorbachev as the Soviet leader in

1985 brought new opportunities for cooperation between the two countries. Within a year of the launch of the first element of Mir, NASA signed a five-year agreement on space cooperation with the Russians. It had taken a little over a decade for the two agencies to consider working together again. In May 1988, the two leaders met for their last Moscow summit and when Gorbachev invited Reagan to see the inside of the Kremlin yard, he apparently suggested to Reagan that they should agree to support a joint mission to Mars. The two visionary leaders understood that to explore space is to do so on the behalf of humanity.

There are currently plans for an international mission back to the Moon and perhaps one day these plans will be expanded to send humans to Mars. The legacy of Mir helped modernize spaceflight to what it is today. It set the stage for the International Space Station to become an international collaborative effort where people from different cultures, speaking different languages all work together to solve problems and perform science in space. In one of the most important partnerships in history, the International Space Station has more than 22 years of operations behind it since launching in 1998 and is expected to persist until at least 2024 — if not longer, until 2028 or even beyond. It is a message for all leaders to consider the benefits of collaboration over the challenge of competition.

Leadership Insights

- Collaborative projects need time for teams to build trust and respect.
- Joint technical projects can be the bridge to working together.
- Space exploration reinforces the importance of international collaboration and globalization.
- Global problems require collaborative solutions.

CHAPTER 10

Normalizing Deviance

"They ate us alive with changes."

— GERALD SMITH

It was a crisp morning at the Kennedy Space Center on January 28, 1986, when the Challenger space shuttle lifted off into the clear blue sky.

"All right,"[1] said mission specialist Judith Resnik, who had been America's second woman in space two years before.

"Here we go," added pilot Michael Smith, who was looking forward to his first venture to space in the coming minutes. Challenger began its roll underneath him. "Go, you mother," he said.

Challenger flew swiftly into the sky, with Smith and Commander Dick Scobee bantering about the wind appearing to buffet the shuttle and how hard it was to see out the windows.

Thirty seconds passed. The crew began routine readouts of speed and altitude for mission control. "Going through 19,000 [feet]," Scobee said. Challenger began its automatic speed reduction to reduce the forces on the shuttle during the moment of maximum dynamic pressure in the atmosphere. "Okay, we're throttling down," Scobee added at 43 seconds after launch.

Underneath the crew's backs, even with the reduced throttling of Challenger's three main engines the acceleration was appreciable. "Feel that mother go!" Smith said, and the crew continued their readouts of speed and altitude.

Moments later, the TV cameras showed something flashing at the bottom of the shuttle. "Go at throttle up," Scobee said.

The flash spread. "Uh-oh," Smith said.

Challenger exploded before everyone's eyes 74 seconds after launch. The crew stopped their callouts. The TV cameras briefly and tragically focused on the parents of Christa McAuliffe, who was on board set to be the first teacher in space.

Now Challenger was in pieces, scattering across the sky. The boosters continued their flight upwards until the range safety officer, following protocol to protect the crowded Space Coast below Challenger, detonated the twin rockets a few seconds later.

"Obviously a major malfunction," remarked a stunned Steve Nesbitt, the NASA launch commentator, of the scene playing out in front of millions.

Flight director Jay Greene was in mission control, out of view of the TV monitors, he said in a 2004 oral history with NASA. But he could see the reactions of astronauts Fred Gregory and Dick Covey, who were serving as capsule communicators for the launch phase. "I saw both their jaws drop," he said. "We got the report that they were tracking multiple pieces, and kept on hoping that some part of the vehicle would come out and everything would have a happy ending, because it was supposed to. It didn't."[2]

A Rough Day

There was faint hope that the crew might have survived the long fall to the Atlantic, he recalled, but it was deemed too unsafe to get the search and rescue teams to the site. Toxic and dangerous debris was still coming down from the sky. Greene followed a procedure which had never been used before during a NASA mission and locked down mission control to gather data, secure communications and have all controllers write out incident reports while their memories were fresh. It took an hour for the debris to clear enough for the search and rescue team to deploy helicopters. There was still no sign of the seven crew members: Dick Scobee, Mike Smith, Christa McAuliffe, Judy Resnik, Ron McNair, Ellison Onizuka or Greg Jarvis.

"[I] was very calm, cool and collected," Greene said. Hours later, the efforts to rescue the crew refocused instead on finding their remains. Greene recalled that he "went home and completely broke down. It was a rough day."

The Challenger disaster was a deadly wake-up call for NASA. There are so many elements that played into the fatal launch decision that entire books have been devoted to analyzing what happened and the myriad problems that accompanied the launch.

Not least among these volumes is the official investigation known as the Rogers Commission. It was led by former U.S. Secretary of State and Attorney General William P. Rogers and included members such as the Nobel Prize–winning physicist Richard Feynman, the famous Air Force pilot Chuck Yeager who broke the speed of sound and NASA astronaut Sally Ride, the first American woman to fly in space.

The major technical cause of the accident happened in the solid rocket boosters. The O-rings, which sealed the aft field joint on the right booster, failed due to the combined effect of a design problem with the cold temperatures the day of the launch.

"Today's tragic event is a reminder that even with the most advanced technology and equipment, and best personnel available, spaceflight is a delicate and dangerous endeavor."
— SEN. BEN NELSON, D-NE

Worse, managers already knew about the design flaw and had seen failures of the primary O-ring that were contained by the secondary in some earlier shuttle flights. NASA and its contractor, Morton Thiokol, had discussed the problem extensively and were aware that a failure in the brittle joint could cause a catastrophe. Ultimately, however, no meaningful changes were made causing the Rogers Commission to conclude the disaster was "an accident rooted in history."

While the natural reaction to disaster is to search for the immediate cause, it is important to start the path to recovery with a full understanding what happened. In her book, *The Challenger Launch Decision*, professor of sociology Diane Vaughan rephrased what had happened as the "normalization of deviance."[3] NASA had continued flying the shuttle at the same time it was evaluating incidents of O-ring damage observed in previous missions.

Culture

"Memos warned of catastrophe. Task forces were formed to try to resolve the O-ring problems," Vaughan said of the environment in 1985. The O-ring issues were essentially unresolved at the time of the Challenger launch and with the aggressive launch schedule in place there was considerable pressure on the NASA team to launch on time.

"After the Challenger tragedy, analysts viewed these expressions of concern as themselves signals of potential danger sufficiently strong that NASA administrators should have halted flight while a new design was implemented," Vaughan added.

Clearly there would need to be a redesign of the booster joint to get the shuttle flying again. The other issue would be addressing the culture. The Rogers Commission found that the night before the launch, contractor Morton Thiokol had several engineers expressing worries about the cold weather affecting the O-rings.[4] They should wait, the engineers urged, but for a number of reasons their concerns did not change the decision to launch. Some of those involved in the higher-level pre-launch meetings were aware of the O-rings' potential for catastrophe but didn't take action.

"Failures in communication . . . resulted in a decision to launch 51-L [Challenger] based on incomplete and sometimes misleading information, a conflict between engineering data and management judgments, and a NASA management structure that permitted internal flight safety problems to bypass key shuttle managers," the commission stated.

Resolving these issues would not be an overnight solution. The shuttle program was grounded for 32 months to fix all the underlying issues, including trying to understand and address the balance between safety and the demanding launch schedule. It was Challenger that showed that, despite all the shuttle could do, it could never be launched every few weeks. The technical complexities of the shuttle and the available resources were incompatible with a launch schedule based on the desire to routinize spaceflight.

In the first few hours after the shuttle's fatal launch, some had the incorrect impression that it was the space shuttle main engine that caused the issue, recalled Gerald Smith, who was then in the office of spaceflight at NASA headquarters in Washington as director of the shuttle main engine project.[5] He had seen the explosion in a conference room and realized, "We've lost it," he said in a 2011 NASA oral history. After only watching five minutes of television coverage, he went back to his office, "convinced that the engine had caused the failure," and spent the day second-guessing all his decisions on the space shuttle.

Rebuilding

Smith was asked to talk to the media, including the *New York Times*, without a lot of knowledge about what had happened. He said he wasn't happy with NASA's request to do so but went ahead anyway and emphasized he was really only doing the interviews for historical purposes. When NASA realized it was the booster at fault, he added, he was then asked to take on the job of overseeing the booster redesign.

In retrospect, it probably wasn't much of a surprise since Smith began working on the space shuttle solid rocket booster (SRB) project back in 1967, working closely with Thiokol as the contractor and NASA to get the shuttle ready for its first flight. Now it would be time to do the same thing all over again.

"Initially the morale was terrible at MSFC [NASA Marshal Space Flight Center] and at Thiokol. They were both being blamed for the accident so the morale was really, really bad. The way I dealt with my own team and with Thiokol is I tried to build one team, because there was a lot of finger-pointing. NASA was blaming Thiokol; Thiokol was blaming NASA for putting pressure on them to fly," Smith said in describing the initial challenge. "I had a picture of the flight crew on Challenger, "Lest We Forget" under it, and that's what would motivate me on weekends and at night. That would keep me going, and we all felt that way. . . . We can't let it happen again."

Smith, of course, was not alone in his efforts. There was a project team at Thiokol working on the matter, along with the National Research Council providing oversight. Every Monday at 6:30 a.m., he would meet with James Thompson, Jr., the project manager for the space shuttle main engine, to brainstorm. "We would meet for about an hour and he wanted to know what's going on and how I was going to solve the myriad of problems I was facing. But part of it was a learning process. I think he was teaching in a lot of ways," Smith recalled. He [Thompson] also wanted to know, "Do you need any help from anybody?" Smith didn't hesitate to tell him when he did.

The solid rocket booster got a thorough redesign. "We reviewed every requirement that existed for the motor and how we verified the requirement. We had to show that every one of these requirements was traceable to either tests or analysis to make sure that we had verified the requirement. This was a major activity that disclosed the many areas in the original certification we had not done." Smith's team was relentlessly working on thousands of issues to ensure success.

The team included rigorous failure modes and effects analysis to identify the many ways a part or subsystem might fail, the consequences of the failure and what might be done to prevent it. Given the naturally risk-averse climate post-Challenger, the engineers submitted thousands of review item discrepancies (RIDs) that had to be addressed before they could return to flight. It was a huge task for the team, but they resolved every one of them before the flight readiness review for the return-to-flight STS-26 mission.

Smith found himself with an added challenge when NASA asked for a fixed-price contract instead of a cost-plus vehicle. The agency was working hard to limit costs on the redesign, perhaps because it was under close scrutiny from Congress to get the shuttle flying again safely and within budget. Smith, however, said his team struggled with the extra parameter. "They ate us alive with changes," he said, "because, first of all, we had not done a good job at documenting the design. As a consequence, we had major cost overruns on the booster."

The team persisted, however, going through the paperwork, verification, testing and all the other steps that were required. They did what it took. "There were [so] many design changes I can't even think of all of them," Smith said, but some of the main ones included the separation system, the separation bolts, the igniter systems and — a new addition — a flight instrumentation package that flew temporarily.

"For the first three or four flights we had extensive instrumentation on the booster to better define the environment and to better understand its operation, and we had not done that originally in the shuttle program," Smith said. "We were faced with installing a lot of instrumentation to define the environments for the boosters which included the separation sequence and the extensive use of cameras."

Success

Remarkably, despite all the design changes, the SRB was ready to fly even before other changes were implemented on the space shuttle itself. Smith estimates that the SRB could have flown about two months earlier than the shuttle actually did, which he considers "rewarding" — not that he wanted to pull those 70–80 hour workweeks again, he added, while working with new management brought in to oversee the issue.

"We had new faces, a new motor project manager — Royce Mitchell, with no propulsion background. We had new people at Thiokol leading the Thiokol effort, even though several of them had solid rocket motor experience. So basically, you had a Marshall team with very little solid rocket motor experience," he recalled of the motor issues. "I had the most, coming from the booster separation world, but essentially we had no propulsion experience. Trying to get the people to believe in themselves again and stop the 'blame game' was difficult. I constantly focused on our challenge to build the best solid rocket motor that's ever been built and getting them to believe [in] that was a major challenge. It took a long time."

Thirty-two months after the loss of Challenger the return-to-flight mission STS-26 launched on September 29, 1988. It was a calm, warm morning when once again the roar of a launch shook the ground at Kennedy Space Center as if to emphasize the resounding success of the return-to-flight effort. For Smith, "It was a great day, because I knew Rick [Hauck] and the entire crew. . . . In fact, later I got a Silver Snoopy from them. I valued very highly the recognition coming from the crew. Rick and his crew were gutsy people to climb aboard Discovery after the Challenger accident." When Hauck and pilot Richard Covey brought Discovery back to a smooth landing at Edwards Air Force Base CAPCOM Blaine Hammond congratulated them on "A great ending to a new beginning."

The consequences of Challenger continued to resonate through the rest of the shuttle program. The Department of Defense, previously a major player in running classified missions and launching satellites on the shuttle, pulled out completely to focus its efforts on expendable boosters. The spectacular spacewalks of the 1980s were sharply curtailed; there would be no more astronauts flying around the Shuttle using jetpacks called Manned Maneuvering Units and no more satellite rescues, outside of the successful missions to the Hubble Space Telescope.

Revisiting the Past

The legacy of Challenger, at least for 17 years, was a safer program. Unfortunately, as time goes by people retire or take on new roles, organizational cultures change and tragically some of the lessons of the past are forgotten. A second vehicle was lost in 2003 — the Columbia space shuttle, killing seven astronauts: Rick Husband, William McCool, Kalpana Chawla, Laurel Clark, Michael Anderson, David Brown and Israel's Ilan Ramon.

The technical cause was different; foam fell from the external tank during launch and struck part of the shuttle, ripping open a hole in the port wing that exposed the inside during reentry. **But many of the culture problems were the same — the pressure of the launch schedule, normalizing deviance, not resolving problems adequately before flight.** The foam problem was a known issue having occurred many times before, but efforts to solve this issue didn't go nearly far enough. It took the loss of a crew to implement a major redesign.

Columbia was a sobering reminder to NASA and its contractors that there is nothing routine about flying in space. There is no room for complacency. The agency implemented new protocols and procedures and tightened up its launch rules even further — such as running most launches in daylight and using video cameras mounted on the shuttle to monitor foam falls during launch.

One of the recommendations of the Columbia Accident Investigation Board (CAIB) called for NASA to develop the capability to inspect and repair damaged tiles in space. Fortunately, it was relatively easy for the Canadian Space Agency contractor MDA to develop an extensible boom with an imaging and sensor package that could attach to the robotic Canadarm, allowing the astronauts to inspect and photograph the undersurface of the orbiter. Tile inspection became one of the primary tasks for the second flight-day of a mission and the imagery sent back to mission control helped the tile teams detect and evaluate any tile damage.

In addition to the orbiter boom sensor system (OBSS), a new maneuver was added to the rendezvous procedures that had the orbiter perform what looked like a backflip as it approached the space station. Astronauts onboard the space station used telephoto lenses to take additional photographs of tile on the undersurface of the orbiter for further assessment of damage. The Space Shuttle Program Office also coordinated the development of a tile repair system that was successfully evaluated during a spacewalk on the return-to-flight STS-114 mission.

The changes worked. Every shuttle launched and landed safely between the restart of missions in 2005 and the last landing in 2011.

Leadership Insights

- Readiness reviews are only effective in controlling risk when identified issues are reported and resolved.
- The lessons for the future are written in the past. Revisiting lessons learned, debriefs and near-miss investigations are all helpful tools in controlling risk.
- Communicating risk requires courage, conviction and clarity when describing the analysis of the underlying data. Recommendations must be clearly stated with verification that the message was heard and understood.
- There is no room for complacency in high-stakes operations.

CHAPTER 11

Rebuilding the Safety Culture

"Spaceflight is still largely experimental.
Routine spaceflight may be a wishful rather
than a realizable goal."
—— SMITHSONIAN INSTITUTION

The investigation of Challenger revealed that NASA's organizational culture was a large contributor to the incident that killed seven astronauts just a minute after lifting off on January 28, 1986. While NASA and its contractors were busy making the shuttle safer through redesigns of the solid rocket booster, the vehicle and its external tank, the agency's culture also needed to change.

"In an instant, one-fourth of the shuttle fleet — the orbiter they had tended for ten missions — was destroyed,"[1] wrote the Smithsonian Institution about the tragedy. "In an instant, the flight rate that had accelerated from two to six to nine missions a year was stalled and the fleet was grounded for more than two years. It was a sobering event for the organization that took pride in preparing the shuttles for flight and an event still remembered with pain."

The tragic loss of the crew was felt around the world, but no one felt it more than the families and friends of the crew and their colleagues at NASA. It was a harsh wake-up call to the agency and contractors when the Rogers Commission unequivocally stated the launch decision of Challenger was "flawed."

"Those who made that decision were unaware of the recent history of problems concerning the O-rings and the joint, and were unaware of the initial written recommendation of the contractor advising against the launch at temperatures below 53 degrees Fahrenheit and the continuing opposition of the engineers at Thiokol after the management reversed its position,"[2] the report stated. "They did not have a clear understanding of Rockwell's concern that it was not safe to launch because of ice on the pad. If the decision-makers had known all the facts, it is highly unlikely that they would have decided to launch 51-L [Challenger] on January 28, 1986."

Testimony from witnesses pointed to several problems within NASA. As the report stated, these problems included "incomplete and sometimes misleading information, a conflict between engineering data and management judgments, and a NASA management structure that permitted internal flight safety problems to bypass key shuttle managers."

Getting the shuttle back to flight would take more than solving technical problems in the booster and shuttle systems. Rather, NASA would need to reinvent its safety culture to prevent such a mistake from happening again. The agency had to rebuild trust with all of its constituents. It had to rebuild its relationship with the astronauts, the administration, congress, the media and the public. Leaders had to reassess how they respond to external pressure by reinforcing their commitment to launch when ready, not necessarily when scheduled. The pressure to launch on time can be a powerful force that is hard to ignore. It must be buffered by a commitment to safety and controlling risk.

The evening before the Challenger launch, NASA was feeling the pressure with CBS Evening News reporting, "Yet another costly, red-faces-all-around space-shuttle-launch delay. This time a bad bolt on the hatch and a bad-weather bolt from the blue are being blamed. What's more, a rescheduled launch for tomorrow doesn't look good either." The host handed the commentary over to a reporter at Kennedy Space Center who continued, "Confidence in NASA's ability to maintain a launch schedule has been rocked by this series of embarrassing technical snafus and weather delays."[3] While it was the agency's unrealistic launch schedule that contributed the biggest launch pressure, even the most seasoned leaders can be susceptible to criticism and media opinion. Shuttle program manager Arnold Aldrich would later comment, "I will say unequivocally today, as I would have said in 1986, that the media could never goad NASA into making any decision regarding a launch. Having said that, I believe such continuous negative scrutiny, to the point of ridicule in the case of the STS-61C events [the mission prior to Challenger that had numerous launch delays], does have a tangible effect on the mood and atmosphere of the conduct of the Mission Management Team activities."[4] In retrospect, "Motives and pressures of cost, schedule, politics, organizational independence/arrogance, pleasing one's customer and media interactions also run strong beneath the surface and affect the manner in which people react and perform."

Restructuring

It is inevitable that the loss of Challenger resulted in significant changes within NASA headquarters and the centers. Operational expertise in senior leadership was critical. Retired shuttle commander Richard Truly left his position as the first commander of the Naval Space Command to return to NASA as the Associate Administrator for Space Flight within weeks of the tragedy. His primary task was overseeing the process of safely returning to flight. He started with the recommendations in the Rogers Commission report and reached out to Robert Crippen, fellow Navy pilot, astronaut and friend for help.

Crippen, a veteran astronaut who flew with Commander John Young on the first space shuttle mission in 1981 and three other missions in 1983 and 1984, had been part of the mishap board trying to figure out the management lessons from Challenger. Crippen moved with a team to Washington for several months, going through the recommendations of the Rogers report.

"We went and interviewed a large number of people that were in various management positions, both within NASA and without, to try to determine how we ought to restructure our management. [We] put together a report with some recommendations,"[5] Crippen said in a 2006 oral history with NASA.

"One of those reports recommended that we needed to get more operational people involved in the Space Shuttle, in the program management of it. When I took that recommendation to Truly, he said, 'Crip, if you really believe that, you'll hang up your flying boots and come take that position.'"

> "What lies behind us and what lies ahead of us are tiny matters compared to what lies within us."
> — RALPH WALDO EMERSON

Crippen thought about it and decided that it mattered more to him that the shuttle fly again safely than having another opportunity to fly in space. He retired from the astronaut corps to take on a new management position, which would be the final authority for launch. It would be based right where the launches happen — Florida's Kennedy Space Center.

"I ended up putting together an office structure both at the KSC, and one here at Johnson [Space Center in Houston] and one at [NASA] Marshall [Space Flight Center, Huntsville, Alabama] to support me in that," Crippen said. Adding these centers to his structure was an important move to enhance collaboration and communication.

Johnson Space Center is the home of human spaceflight. It is where the astronaut office is located as well as most of the management structure that manages the astronauts and missions. The support of that center would be essential to keep the astronaut office involved in launch decisions. Marshall was where rocket development was centered; as the failure of the solid rocket booster had been the proximate cause of Challenger's loss, it would be essential to keep those teams in the loop for safety considerations.

Crippen reassessed all elements of the launch process and focused on who would chair the flight readiness reviews, particularly at the L-minus-2 review that took place two days before launch. This was a critical event where a mission usually got a final go/no-go decision before the start of the final flight readiness procedures. He reviewed who would sit in the "firing room," the launch control facility at Kennedy Space Center located a little over three miles (5.6 km) away from the launch pad. The team also came up with stricter "launch commit" criteria, particularly those for weather.

Members of the public and media often complained about shuttle launches being delayed for weather. But these safety criteria were crucial when human lives were involved, Crippen's group decided.

Truly worked with the team at NASA headquarters to implement several key management changes in the wake of the Challenger space shuttle disaster, summarized in a Safety Division report from February 1988:

- Appointing a new NASA associate administrator for safety, reliability, maintainability and quality assurance.
- Creating a pathway to clearly define who is responsible for what, especially in the case of dual roles that were poorly defined before Challenger.

- Augmenting the skills and resources for safety, reliability, maintainability and quality assurance through augmenting the budget and maintaining training of personnel (an important thing to keep an eye on as managers leave).
- Creating a procedure for deviation and waivers.
- Ensuring that program management is skilled and motivated, and that critical knowledge of the program is passed from employees to management.
- Maintaining an effective problem reporting and corrective action system.
- Focusing on trend analysis, safety risk assessment and a good flight readiness review system.
- Maintaining several systems related to safety: assurance information, engineering changes, crew safety planning, contractual safety requirements and program planning, as well as running contractor selection processes with an emphasis on safety.

He clearly had his hands full addressing the many technical and organizational issues that had to be resolved to ensure the safe return to flight. At the same time, he was managing more recommendations from other advisory groups concerned about the agency's organizational structure and decision making. It was a very difficult time for the agency.

The finding of "flawed decision making" permeated NASA leadership leading to conflict and finger pointing. The leadership and workforce were initially reeling from the tragedy, grieving the loss of the crew and struggling with guilt wondering what they could have done to have prevented the disaster. These reactions were soon followed by a lot of resentment, depression, exhaustion and hostility. Truly remarked at the time, "I have a lot to do."

A natural leader, Truly understood the importance of communication, engagement and teamwork to achieve success. He developed "orderly, conservative and safe" criteria for return to flight and he presented his plans to over one thousand employees at Johnson Space Center ensuring they were also televised to other NASA centers. Looking out at the assembled group in the Teague auditorium he said, "The business of flying in space is a bold business. We cannot print enough money to make it totally risk-free. But we are certainly going to correct any mistakes we may have made in the past, and we are going to get it going again just as soon as we can under these guidelines." The *New York Times* later reported that "his upbeat words appeared to be meant to lift spirits at the beleaguered agency and to turn the staff's eyes forward to the shuttle's future."

A new position for a director of the Space Shuttle program was created at NASA headquarters and Aldrich moved from Johnson Space Center to Washington to fill the role. In addition to the changes outlined in the Rogers report, Aldrich determined, "After we finally got through the failure analysis and sorting out what really happened with Challenger . . . I realized that [the SRB joint] was not, maybe, the only weak system in the Space Shuttle. I set up a series of formal reviews looking at all aspects of the Space Shuttle for risk management, for weaknesses."[6]

Using an approach similar to the one he had seen Low use after the Apollo fire, he convened a series of change board meetings to "look at everything and bring those things in and talk about them and prioritize them." He and his team made over 250 changes to different elements of the Space Shuttle. Critical changes and a number of others were implemented for the return-to-flight mission while additional changes brought onboard later. "Of all those changes, only about five of them dealt with the solid rocket motors. They were big changes, but [those that were] the cause of the Challenger that we had to fix [were] only about five of them."

There were consequences from this renewed emphasis on safety, Aldrich wrote in a paper for NASA in 2008. He noted, "The combination of the impact of the Challenger accident and these in-depth program technical reassessments and changes had caused a predictable amount of additional conservatism and caution across the entire Space Shuttle program community."[7]

One of the key short-term consequences was the impact on the space station Freedom program, which was already facing over-runs, design changes and an unrealistic budget as the Reagan administration and NASA tried mightily to get this station to orbit.

In Aldrich's estimation, Freedom could have flown if Challenger had not happened. He did not go into the details in his paper beyond citing the cost overruns and other delays, but the loss of Challenger delayed the launching of any other spacecraft for about 32 months. The Space Shuttle program was supposed to ship up pieces of the space station, but that clearly couldn't happen when the program was grounded. And by the late 1980s, when the shuttle was flying again, there was a backlog of missions that needed to be dealt with.

Another consequence Aldrich noted was the pullback of Department of Defense. The DOD had been an important ally in the early years of the program, with the payload bay of the shuttle having been redesigned to make it possible to haul up large DOD satellites. The loss of Challenger delayed several DOD missions and created a backlog that began to have national security implications. They would have to find another solution and the DOD made a permanent and irrevocable shift, eventually pulling out altogether.

The shuttle was originally intended to be used as a reusable launch vehicle for more cost-effective commercial and scientific missions. Unfortunately, the complexity of the shuttle design, a more realistic revised flight rate and the evident risk significantly diminished the opportunities for DOD launches and commercial payloads. Ultimately, the shuttle never ended up being a commercially viable vehicle, although it did very well in its mission to build the International Space Station and provide a research capability in space.

Challenger was a sobering reminder that space would never be a routine exercise. NASA had been implementing programs in the preceding months and years that brought people with a minimum of training into space, under the payload specialist designation. It was an excellent program that allowed international astronauts to participate in NASA missions, including the Europeans, the Japanese and the Canadians.

"What was learned from the Challenger tragedy? Some of the lessons were obvious in retrospect, and they were basic principles of rocket engineering, but they reminded us **how easy it is to become comfortable, even careless, with responsibility.** Spaceflight is inherently risky, and there are no shortcuts to the management of risk," the Smithsonian said of lessons learned.

"**Vigilance is the price of safety, and vigilance cannot be relaxed. Pay attention if something isn't right; it may be telling you something important. Communicate clearly and be disciplined in decision making.** Spaceflight is still largely experimental. Routine spaceflight may be a wishful rather than a realizable goal."

Leadership Insights

- Organizational culture is a critical determinant of success. While there are many measures of performance, it is very difficult to detect subtle changes in culture that affect operations.
- Build risk management into program management.
- Rigorous configuration review and control is critical.

CHAPTER 12

What Can We Afford?

"No Bucks, no Buck Rogers."

—— TOM WOLFE

Congressional concerns over NASA budgetary overruns have been a worry for most of the agency's history. They have led to the shifting priorities, cancellation of programs and a change in leadership. Faced with the challenge of balancing the national budget, President Nixon canceled the Apollo program after the Apollo 17 mission and cut NASA's budget. George Low commented, "His general reaction to us was he wanted to move forward . . . with a meaningful program. He did not want to be the president that stopped all the wonderful things that had happened."[1] However, the competitive pressure of the previous decade had ended. NASA had achieved the goal of landing humans on the Moon and returning them safely to the Earth by the end of the decade and there were other pressing issues facing government. What would be next for the agency?

In its first decade, NASA built a strong competency-based foundation to support the Apollo vision. That experience had shown that bold visions require a unique combination of technical competency, teamwork and commitment. As director of NACA in the late 1950s, Hugh Dryden challenged his engineers to be "bold and imaginative," declaring a long-range vision: "The goal of the program should be the development of manned satellites and travel of man to the Moon and nearby planets." That vision was embraced by the newly founded NASA team, including many who would later take on significant roles in the agency — James Webb, Bob Gilruth, Max Faget, Christopher Kraft, Wernher von Braun and George Low — whose leadership built the teams that succeeded with the Mercury, Gemini and Apollo programs. Within its first 10 years, NASA had achieved a goal that many individuals thought was impossible. NASA had to develop a new vision for its future.

Station or Shuttle

The development of the space shuttle was a critical step forward in NASA's plans post-Apollo, representing a shift from building a space station based on the Skylab experience to developing a reusable vehicle that could be used to perform a number of tasks and open the door to the commercial utilization of space. "A space station without [a] shuttle, without a good transportation system to build it and to make use of it, makes no sense at all. A shuttle without a space station does, and so we decided to go for the shuttle first." Either way, Low had a plan. Implementing that plan required successfully navigating the many political forces governing NASA's future while managing the ever-shrinking post-Apollo budgets.

In the opinion of space historian Roger Launius, "If I were to point to the most important thing that George Low ever did, it was rescuing the agency after it had crashed and burned in the Nixon White House. . . . Low's philosophy was to strike a balance between the practical and the visionary and to move the agency forward by investing in future technologies and capabilities rather than remaining invested in the old and outdated."[2] Low succeeded, securing funding to develop the space shuttle in preparation for what later would become known as Space Station Freedom. Lunar return and possible missions to Mars faded into the background and President Nixon announced the approval of the shuttle program with the new NASA Administrator James Fletcher in early 1972. Within a decade, John Young and Bob Crippen would fly the first shuttle mission STS-1, on April 12, 1981.

To leverage the capability of the shuttle and create a new foreign policy initiative, in his 1984 State of the Union address, President Reagan directed NASA to build a space station within the decade and invite other countries to participate. For the Reagan administration this was an important opportunity to demonstrate to the Russians that America "want[s] our friends to help us meet these challenges. . . . so we can strengthen peace, build prosperity and expand freedom for all who share our goals." Appropriately, the space station would be called "Freedom," and it was clear to NASA that the best way forward was one of international collaboration. A formal agreement was finally reached in 1988 with Europe, Canada and Japan committed to the partnership. Unfortunately, large government projects often take on a life of their own with cost overruns and scheduling delays, and such was the fate of the space station Freedom. Costs rose without moving past the design stage, and despite the involvement of the other partners, little tangible progress was being made.

Shortly after the election of George H.W. Bush, who followed President Reagan in 1989, James Fletcher resigned. The Bush team appointed Vice-Admiral Richard Truly as the next NASA administrator tasked with setting new goals for the agency. To provide oversight, Vice President Dan Quayle became chair of a newer version of the original National Space Council with administration staffer Mark Albrecht to help oversee the preparation of a compelling vision for human spaceflight that would build on the space policy of the Reagan administration. Albrecht had been influenced by NASA's achievements in the Apollo era. "As a person who grew up in the sixties . . . I was an absolute fan of the space program, watched every Mercury, Gemini, Apollo launch just like every other person in my generation," and he was excited to be part of creating the new opportunities in space. In celebration of the 20th anniversary of the Apollo 11 lunar landing, President Bush announced a new Space Exploration Initiative that would return humans to the Moon and go on to Mars using the planned Space Station Freedom as a stepping stone to learn about the many issues associated with humans living and working in space for long periods of time. This bold plan would require additional funding beyond that already committed to Freedom.

Truly asked JSC Director Aaron Cohen to undertake a 90-day feasibility study to determine the potential cost of the new plan. Cohen's group of internal NASA experts estimated costs of up to $540 billion over 20 to 30 years to build a space station, return to the Moon and go on to Mars. This was clearly unsustainable in the fiscal climate at that time, given the exponential growth of the Freedom program's costs to close to five times its initial cost of $8 billion to an estimated $38.5 billion within the year. The Space Council called for an external review by a newly formed Advisory Committee on the Future of the United States Space Program chaired by Norman Augustine, the CEO of Martin Marietta.

A Growing Bureaucracy

An Augustine committee[3] was asked "to advise the NASA administrator on overall approaches NASA management can use to implement the U.S. space program for the coming decades." Their report acknowledged the many challenges NASA faced: "The source of this criticism ranges from concern over technical capability to the complexity of major space projects; from the ability to estimate and control costs to the growth of bureaucracy; and from a perceived lack of an overall space plan to an alleged institutional resistance to new ideas and change. The failure of the Challenger, the recent hydrogen leaks on several space shuttle orbiters, the spherical aberration problem encountered with the Hubble Space Telescope, and various launch processing errors such as a work platform left in an engine compartment and discovered during launch preparations, have all heightened this dissatisfaction." The Committee did acknowledge that "NASA, and only NASA, realistically possesses the essential critical mass of knowledge and expertise upon which the nation's civil space program can be sustained — and that the task at hand is therefore for NASA to focus on making the self-improvements that gird this responsibility."

Truly took on the challenge of addressing the Commission's concerns and implementing its recommendations but "it was felt [by some officials] that he was captive of his bureaucracy and incapable of making the changes, the reforms, the administration wanted."4 Truly had played a major role in the shuttle return to flight activities following the loss of Challenger, and there was widespread recognition of his efforts. John Logsdon, a space policy analyst at George Washington University and adviser to the space council, said Truly "did an extremely valuable job in getting the shuttles flying again, and restoring a sense of integrity to the agency," though subsequently adding, "Truly's vision of the future was not compatible with the realities of the world."

Truly had also asked Apollo-Soyuz Test Project veteran General Tom Stafford to lead a comprehensive study that would become known as the Synthesis Group. While the 1990 Augustine report called for "redesigning the Space Station Freedom to lessen complexity and reduce cost, taking whatever time may be required to do this thoroughly and innovatively," it was General Stafford's Synthesis Group that would find the path forward. The individual who played a critical role on the journey was former Air Force pilot George W.S. Abbey.

Richard Truly would later say that "the real book about the manned space program would be a book about George Abbey."[5] Described as "one of the most elusive and controversial figures in NASA's history," he had played a key role working with George Low in rebuilding the Apollo program after the Apollo 1 fire. Abbey recalled, "I had the good fortune in Apollo to go to work for George Low as his technical assistant and as such I got involved in every aspect of the Apollo program. After we had the Apollo fire and George Low came in as the program manager, he was looking for a way to bring all the elements in the center together, the designers, the operational people, the astronauts and the scientific people and have them work together. I suggested that he create a configuration control board (CCB) that would include all these elements as current members of the board as well as the directors of each of the contractor organizations."

The Apollo CCB would prove critical to the success of the Apollo program and the close working relationship and mentoring between one of NASA's outstanding leaders and his assistant would ultimately play a critical role in NASA's future. Apollo astronaut Ken Mattingly commented, "The two Georges were a remarkable team in that George Low did everything in public and did all of the formal stuff and wrote memos and gave directions, and everything he did was a matter of record. Abbey, on the other hand, was intimately involved in every one of those things, every conversation, but he also had this network of working people. He knew all the troops and all the buildings, and he'd wander around and just talk to people and bring all of that stuff back. He knew what George Low was concerned about, the kind of questions, and so he would bring that stuff back, and informally, just no attribution, he would just make sure that George Low was aware of everything going on as perceived from the bottom of the barrel, as well as the reporting structure that officially brought things in."[6] While George Low was well known for his relentless commitment to focusing on the details of an engineering project, George Abbey gave him those details.

Disagreement and Discourse

Coming from a traditional military command structure in the Air Force, Abbey was surprised by his first encounters with NASA teams during his work on the Dyna-Soar space plane project in the early 1960s. "I was sent to participate in a conference at Langley Research Center, which was going to address the configuration of various re-entry vehicles — whether you would have a winged vehicle or whether you have a capsule. . . . I was sent as the Air Force representative the week before the conference started. There were people from all the NASA centers, also small groups from the contractors and the Air Force but mostly from the NASA centers. They had each person present the paper that he was going to give the next week of the conference. He would stand up and go through the presentation and they had a moderator, so when he finished his presentation the moderator would say, go back to the first chart. So they would bring the first chart up, and then the audience would attack him about practically every data point on his chart. It got to be a very violent meeting. People were calling each other names and disagreeing wildly about concepts and data. The moderator would listen to all this and then slam his gavel down and then he'd say, 'Gentlemen, these are the points to change in the chart.' That went on every day, but at the end of the day they would all assemble at the gym and set up kegs of beer outside. After this day-long meeting with people yelling and screaming at each other, we would all get together and have a beer. It was like they were long-lost buddies, like they had never yelled or screamed at each other. I suddenly realized, really, that in the culture at NASA, and I saw it when I came to Johnson the first time as well, there was a great openness and people said what they thought. You had an opportunity to really bring your opinions forward, and you had the chance to defend them. They would get addressed and I found that to be culturally very unique and very special and I think it led to the successes you saw in NASA all through the early phases of the program."[7]

His new role model, George Low, embraced the Langley culture as chair of the Apollo CCB. Abbey remembered many loud arguments between the various NASA people and contractors. Low's method was to let the arguments play out then slam down the gavel and state the resolution. Low believed it was important for people to speak up at team meetings. Between themselves, team members shared, **"Don't be afraid to disagree, but don't state an opinion that you can't back up."**[8] Abbey learned an important lesson from Low: "It's important that you never want to put yourself in a situation where you're surrounding yourself with people that agree with you all the time. **I think it's important to have people that work for you that tell you when you don't know what you're talking about.** That's an important element of leadership and the leaders I grew up with managed to surround themselves with people that didn't always agree with them because they wanted to hear different opinions. I wanted to hear those opinions. I would have to admit . . . that made me think about whether what I was going to decide was the right answer or not and very often, it wasn't the right answer. They were right. **So, surround yourself with people that are well informed, have good backgrounds and are willing to speak up. You don't want people around you that are going to agree with you all the time.** You're headed for trouble that way."[9]

Continuous Learning

How to lead peak-performing teams would subsequently become part of the astronaut expeditionary behavior training for long-duration missions to the International Space Station. Those astronauts are sent, often with their fellow crew members, to the National School of Outdoor Leadership to learn about high-stakes leadership, teamwork and followership, which involves "supporting the designated leader and the group's goals by giving input, respecting the plan, and staying engaged."[10]

> "One of the criteria for national leadership should therefore be a talent for understanding, encouraging, and making constructive use of vigorous criticism." — CARL SAGAN

Low gave Abbey the responsibility of developing an agenda for the CCB meeting each week that was distributed on a Monday to the program and the contractors. Abbey recalls, "They would have from Monday until Friday to get ready to present and discuss the issue and he [Low] insisted on having the individual designer who was responsible for that subsystem speak to the topic from the NASA side with the contractor present. He [Low] insisted on having the leadership of the contractor and each of the organizations at the center in attendance at that meeting so we would get a decision right there and were able to document that decision and get it out on Saturday to the program. It brought all the elements of the program together and gave a voice to not only the NASA elements but also the contractors."[11] Everyone had the opportunity to speak up, decisions were made, and they were immediately communicated to those responsible for implementation.

Having worked with Abbey in the ASTP, General Stafford immediately added Abbey to the group as the senior NASA representative, a similar role to the one he had so effectively fulfilled working with Low on the Apollo CCB. Abbey's network was wide, and the Synthesis Group was quickly constituted with key influencers in the aerospace sector. Abbey had always possessed a unique ability to recognize highly competent individuals, a skill that was critical in his former role as the head of the Flight Crew Operations Directorate at Johnson Space Center. For over a decade, in his roles as director of flight operations and then as director of the Flight Crew Operations Directorate, Abbey was involved in hiring and assigning astronauts for the shuttle missions. In his youth, he had learned the importance of doing what was right. In the case of astronaut assignments, **hiring the best candidates and assigning them to missions that matched their skill set was something he wanted to always get right.**

Achieving Consensus

The Synthesis Group immediately got to work with a sense of urgency to help clarify the vision and priorities of the Bush administration's space exploration initiative. Jeff Bingham, the space staffer for Senator Jake Garn who had flown on the space shuttle Discovery mission STS-51D, joined the group and described the meetings as a "nonstop succession of hour-long sessions with experts in every aspect [of human spaceflight]." With Abbey's guidance and Stafford's support, the 12-hour days were filled with presentations from, in Bingham's words, "every rocket designer, every propulsion system [expert]; we had people talk about the psychology of long-duration spaceflight. It was just nonstop and endless, and we divided up in teams and came up with four different architectures."[12] The Synthesis Group members were kept busy preparing the material for the formal meetings. Albrecht recalled that "the real Synthesis Group probably only met three or four times and everything they saw came from us, and George." Within the year the 93-member group and 35 technical advisers synthesized information from 126 presentations to produce a comprehensive report articulating specific recommendations for Mars exploration, a science emphasis for the Moon and Mars, a plan to go to the Moon "to Stay," followed by Mars Exploration and Space Resource Utilization.

Leveraging proven strategies, the group's meetings were often followed by informal beer calls. Abbey would later reflect: "The important thing for the NASA leadership is to have access to the knowledge and understanding that comes from working at the lower levels and developing that experience base and getting a good understanding of the design activities, testing activities and operational activities."[13] Whether at the lower levels or the most senior levels of aerospace leadership, a key element of his approach was asking questions and listening. There was no better setting than informally getting together in groups after the hard work was done to sit and chat about how things really were.

Vespers

These sessions were so effective Abbey later implemented more formal Friday evening sessions called Vespers where invited guest presenters were informally asked to share their thoughts and perspectives on the topic at hand. These sessions were of such value that Stafford once skipped a paid board meeting in California to avoid missing a meeting. True to its title "America at the Threshold," the breadth and depth of the Synthesis Group report was so comprehensive that it could be used as a template for NASA's current vision of sending humans back to the Moon. The group also provided a list of guidelines and pitfalls for future leaders based on 40 years of collected wisdom of spaceflight leaders. **The lessons for the future are written in the past** and these aphorisms are as relevant for spaceflight in the 21st century as they were then.

With the Synthesis Group report adding tactical recommendations that would support the Space Exploration Initiative, NASA was still struggling to contain the ever-growing costs of Space Station Freedom. Under Quayle's leadership, the Space Council realized a new approach was required to successfully implement the vision of two administrations to build Space Station Freedom. Mark Albrecht, familiar with Abbey's role in the Synthesis Group, decided that he could use Abbey on the Space Council. Truly objected as he knew how valuable Abbey was to the agency. Finally Albrecht asked Truly, "Should I have the vice president call you to make the request?"[14] Truly relented and Abbey was excited to take on this new assignment.

Flimflam

Albrecht recalled, "We knew — and I think Truly knew — that with George at the Space Council, the White House couldn't be flimflammed by NASA. Talking to him was like having a decoder ring for NASA. . . . I was fairly animated, and I would be waving my arms and saying, 'What the fuck are they doing?' And he would shuffle and shift and mumble, 'This is what's going on.' And he was always right. He was loyal and he was smart." Former space shuttle commander Robert "Hoot" Gibson said of Abbey, "So many of the [NASA] administrators or center directors were exceptional technical people, and exceptional scientific people. But I think no one ever had the grasp that George did of the political forces in play that had a big effect on NASA's future and NASA's success."[15]

Congress subsequently directed a redesign of the space station and in March 1991 NASA submitted a new proposal for $30 billion. To many in Congress, this proposal further emphasized the growing concern that NASA spending was out of this world. Later that year a House floor vote on the fate of the program would be held. Without a space station, there would be little purpose for the space shuttle. Without a space shuttle, the future NASA plans for human spaceflight seemed to be hanging in the balance. To Albrecht and others, it was time for NASA to undergo significant changes. In the world of large government organizations that would mean a change of leadership.

Albrecht knew Dan Goldin. "I was privileged to be in a position to have come across him in the course of my business in the Space Council prior to his appointment as administrator, and he had impressed me . . . from [our] very first encounter . . . with his ability to visualize where we were trying to go and to understand fundamentally what some of the impediments were." Goldin saw many of the challenges NASA was facing from an industry point of view: "He had strong vision in his job and industry about how to move beyond and how to facilitate a growth in the commercial space business and in supporting government institutions that was so consistent and so clearly motivated by the things that I saw."[16] Albrecht felt Dan Goldin would be the ideal candidate to replace Truly.

Leadership Insights

- Don't state an opinion that you can't back up. The best decisions come through respectful discourse, disagreement and listening.
- Build teams with talented people who are well informed, have good backgrounds and are willing to speak up and listen.
- Have access to the knowledge and understanding that comes from working at all levels in the organization.

CHAPTER 13

Faster, Better, Cheaper

"Frankly, I think NASA's been pretty
rock-and-roll under his administration."
—— SEN. BARBARA A. MIKULSKI (D-MD)

It was known as the summer of discontent. The start of the last decade of the millennium was particularly difficult for NASA Administrator Richard Truly. The shuttle fleet was grounded following a series of leaks associated with the hydrogen disconnect valves that were an integral part of the fuel supply to the space shuttle's main engines. The Hubble Space Telescope had launched in April and soon after it was discovered that the telescope's mirror had been ground incorrectly, resulting in out-of-focus images from a $1.5 billion scientific instrument that was initially lauded as the eighth wonder of the world.[1] The failure was widely covered by the media with numerous references to it as the myopic telescope, the white elephant and a techno-turkey. Late-night talk show host David Letterman summed up national sentiment in a now infamous segment called "Lost in Space: The Top 10 Hubble Telescope Excuses."[2] If that were not enough, the Space Station Freedom project, President Reagan's visionary space station that was to be built as a collaborative multinational research platform in low Earth orbit, was over budget, behind schedule, the subject of congressional scrutiny and close to cancellation. Truly had a number of challenges. He initially had the support of President Bush who said he had "great confidence" in Truly, further commenting, "It's such a complex organization that it is appropriate that the administrator now call on the best minds he can find to see how we're going to meet these goals."

The president's Space Exploration Initiative (SEI) and its bold vision for NASA to send humans back to the Moon then on to Mars was visionary, inspirational and would ensure sustained American leadership in space exploration. Dan Goldin recalls that President Bush "was one of the most space friendly, visionary presidents,"[3] a perspective shared by the Space Council. Since the Commercial Space Launch Act of 1984, there had been a growing shift in the vision for government agencies like NASA away from being the lead in proposing, designing and bringing to fruition large space exploration projects to providing support for the development and growth of commercial space initiatives. This necessitated a more collaborative way of working with private sector companies while moving away from large, big-budget, complex projects.

Challenges

Goldin, with his private sector background, would become the next NASA administrator tasked with creating a sustainable future for American space exploration. Time was not on his side as he would have to deal with the list of issues noted in the Augustine report and a host of other challenges that needed to be addressed. He was ready for the task.

Goldin began his career at NASA in 1962 at the Langley Research Center in Cleveland, Ohio, working on electric propulsion systems for human spaceflight. He subsequently left to work at the TRW Space and Technology group, eventually becoming vice president and general manager leading advanced communication spacecraft projects. He and many others felt that NASA was becoming overly bureaucratic. "The NASA budget kept growing, growing, growing and growing. The management techniques that they were using, the tools that they were using, for the most part, were obsolete."[4] Goldin was excited about the opportunity to come back to NASA. He realized significant change was required and understood that **effective leadership is not a popularity contest.** Described as "strong-willed, confrontational and decisive,"[5] he quickly became known for his credo "faster, better, cheaper." His was a mandate for change and he became known for the changes he was making. His approach was popular for many outside the agency, but the change in philosophy did not sit well with some of the longstanding career civil servants within the agency.

Change Is Difficult

Truly left the agency and Goldin, who would become known as one of the most influential administrators in the agency's history, took on the challenge in March 1992 saying NASA would have to "do more with less." It was clear change was needed. He turned to General Tom Stafford and Albrecht for advice. They suggested George Abbey and Goldin agreed. Based on his work with the Synthesis Group, Abbey recalls that at that time "I was more of the opinion we should take advantage of the Mir program and the Russian experience and move forward to go back to the Moon" than committing to build a space station in low Earth orbit. Whatever the vision, if programmatic costs could not be managed there was a risk they'd be canceled.

"Effective leadership is not about making speeches or being liked; leadership is defined by results not attributes."
— PETER DRUCKER

"I know I made everyone crazy," Goldin said, describing his early tenure at the agency. While addressing the many changes he envisioned, he felt that "[the] safety of human beings was the number one objective." He "decided that safety — first of those in highest risk; the astronauts, then other humans on the ground, then high-risk, high-value robotic systems was the prioritization. And I just drove, drove, drove, drove"[6] to achieve those priorities. Many organizations speak of the importance of safety but fail to incorporate safety as a critical element of their culture. Goldin immediately recognized his role in ensuring operational safety and made changes to his senior team, appointing former Marine General Jack Dailey as the deputy administrator and George Abbey as a special assistant to the administrator to underscore that commitment. "If the leader of the organization doesn't take personal responsibility and is not willing to risk their reputation and career on that you have no safety system. I had that problem, not just with NASA, but I have had it with other functions of government and in industry. . . . But I always think about accountability."

Accountability is a key element of organizational culture in controlling risk.[7] Both Dailey and Abbey had significant first-hand operational leadership experience, which Goldin used in reshaping the agency. Veteran space shuttle commander and senior executive Jim Wetherbee noted, "It's critically important that any leader understands the culture in the organization when he or she arrives." Goldin listened to both Abbey and Dailey, using their recommendations in his decisions to reduce the size of the workforce while reshaping the program to be more cost-effective. His leadership, trust and willingness to listen to technical experts resulted in 59 space shuttle missions without incident during his tenure as NASA administrator.

Building Trust

There were two key objectives Goldin had to achieve early in his tenure while concurrently dealing with the Hubble telescope repair, ongoing shuttle issues and space station challenges. He had to regain the trust of the White House and Congress, and he had to bring the NASA budget under control. "When I arrived, the Augustine report had been written recently saying the NASA budget was going to go from $14.3 billion to $25 billion over ten years," Goldin recalled. "During my confirmation preparations, I had a conversation with Senator Fritz Hollings. . . . He went to the board and he drew a curve. And he said, here's your budget here. . . . He said, if you want this job, make a flat budget. The budget is too high. He said, Hubble is blind, Galileo is a mute. It can't speak because the antenna didn't deploy — that's a multibillion-dollar spacecraft on the way to Jupiter. The shuttle is grounded with all sorts of problems. The space station spent its budget and they didn't do a damn thing. There was one, oh yeah, and the weather satellites. The geostationary weather satellites are dead. Hurricane season is coming. And then NASA has no vision. It's out of touch. Do you really want this job? I said, Yeah, I'll take the flak. And I know exactly how to fix it."[8]

"So, I knew right away, one-third of the staff is going to go. . . . You're [going to have to] get more productive. I got an agreement to get a buyout because I didn't want people to suffer. I wanted voluntary leave . . . as it turned out, NASA led the pack to get the buyout. . . . We didn't have one forced layoff of a NASA employee and we went down from roughly 25,000 to 17,000. You have to see the big picture. You have to see where you were and how did you get to where you are? And you can't go beat up on people."[9] Goldin wanted to change the culture of the agency to embrace the faster, better, cheaper approach he had developed at TRW through the use of innovative technologies and new approaches to managing large projects. "When you're in a transformational job, you need to do two things. You need to make an assessment of where the organization is and you go back in its history, and you say, how did it get here? You don't just go in and slash and burn . . . you say, where do I need to go? You need to visualize the future. Then you work backwards to where you are today." Goldin wanted to transform the agency with new programs and new ideas that would propel the agency into the 21st century.

Visionary Leadership

To do so he would have to be reappointed by the Clinton administration. Less than a year after starting, with agency morale at a low point struggling with his vision for change, Dan Goldin was at risk of becoming a one-year administrator. Fortunately, the new administration brought a mandate for change. In a short time, Goldin "has extensively reorganized the NASA management chart. Reorganizations . . . often happen in government, but the resulting product is usually even more unwieldy than before, since no one ever loses authority while various new layers of paper-pushers are added. In Goldin's case, several entrenched bureaus actually have been zapped."[10] Change can be perceived by some as a threat: "Whenever senior civil servants start complaining about low morale, what they mean is that someone is pressuring them to change their ways. In that sense, nothing could be better for NASA than a little bad morale. The cause of the complaining is that fundamental changes at NASA are, at long last, after much talk, finally . . . being made. For nearly a decade, NASA has staged elaborate rituals of sham reform."[11] This was not a sham reform. In 10 months, Goldin had "pushed his staff to rethink how the agency performs its basic mission, to cut waste from operations and to change the way NASA does business with private contractors. He . . . made at least 35 personnel changes in key leadership posts at NASA headquarters, promoting new favorites and dumping or reassigning others who had influential posts. . . . He has split up old NASA offices and created new ones to focus more attention on areas such as aeronautics, planetary science and the launching of satellites to monitor Earth's environment."[12] He had made the most of his mandate for change and his reappointment by the Clinton administration enabled him to continue to transform the agency.

Goldin's vision for the future was based on a desire to build on the achievements of the early days of space exploration by encouraging staff to "dream again," to think broadly about the agency's vision for the future. To some, one of Goldin's first actions seemed trivial at the time but became an important element of rebuilding the NASA culture. In his first month as administrator, Goldin banned the "worm," the new, futuristic NASA logo designed in the 1970s, to reinstate the original NASA "meatball" logo. Shortly after his confirmation, Goldin had begun visiting NASA centers to learn more about the challenges each were facing. When he landed at Langley his plane taxied towards a hangar that proudly displayed the new worm logo and the original meatball logo above the hangar doors. Many at NASA did not like the new logo, and when Goldin asked George Abbey and Langley Director Paul Holloway what could be done to improve sagging morale at the Centers, Abbey suggested Goldin restore the meatball, Holloway agreed and Goldin made the announcement at the end of his speech stating "The spirit of the past is alive and well."[13] For many, "The NASA emblem taps into a sense of universal wonderment about what is good in this world because it was one of the few entities that successfully allowed us to leave it."[14] This was not a comment about graphic design, it was a commitment to ground the NASA culture in a compelling vision and the values of competence, integrity and trust that were critical to the success of the Mercury, Gemini and Apollo programs.

Transformation

Goldin continued to transform the agency and within five years, NASA had accomplished the risky but successful Hubble Space Telescope repair mission, begun commercializing the shuttle program, implemented a new plan to redesign the space station in collaboration with Russia and the original international partners and had significantly reduced the workforce. "Mr. Goldin has a bold personality," Sen. Barbara A. Mikulski (D-MD), who then chaired a key NASA funding subcommittee, said. "But frankly I think NASA's been pretty rock-and-roll under his administration."[15] John Young, Apollo 16 commander and commander of the first space shuttle flight STS-1, said "that despite his rough edges, Goldin's heart is in the right place." Apollo 10 and Apollo-Soyuz commander Thomas Stafford considers Goldin the best NASA administrator since Jim Webb, who laid the foundation for Apollo during the 1960s.[16] Although many challenges remained, Goldin was able to shift his focus from human spaceflight to robotic missions to explore Mars and the Earth's solar system while relying on his trusted advisers, his leadership team and center directors to continue implementing the new strategic plans.

Goldin was finally able to focus on his desire to explore Mars. "I made a pact with myself, so I can't back off. . . . I accept the fact that maybe there may be a higher priority for human beings to go to places other than Mars. But my deepest dream is that in my lifetime I will [in] some way be responsible for a mission to Mars. It would be the next noble thing that we could do as a society."[17] He created the Origins program to understand the beginnings and destiny of the universe while looking for conditions that might support life. This would eventually lead to the creation of the National Astrobiology Institute initially directed by Scott Hubbard before the appointment of Baruch Blumberg, a Nobel Prize–winning biologist and physician, as the first NAI director. Goldin also committed NASA to using the faster, better, cheaper strategy to launch a series of orbiting spacecraft and landers that would go to Mars every two years, a plan that was endorsed by the Clinton administration. President Clinton started his second term in January 1997 and there was little question he would retain Dan Goldin as the NASA administrator.

Restored

That summer the Mars Pathfinder mission, designed using Goldin's faster, better, cheaper strategy, would land on Mars. In addition to using the new management strategy, the mission was also a "proof-of-concept" for various technologies, such as airbag-mediated touchdown and automated obstacle avoidance, both later exploited by the Mars Exploration Rover mission. Pathfinder was the $171-million successor to the 1992, $1-billion Mars Observer mission, bringing credibility to Goldin's faster, better, cheaper management strategy. Not only was Pathfinder better and cheaper, but NASA also leveraged the internet by creating a website enabling people worldwide to follow along as the Sojourner roving vehicle explored the Martian surface. In addition to scientific objectives, the new possibilities for education and outreach paralleled those of human spaceflight.

Goldin had successfully led NASA through a period of significant change. Reflecting on the challenges Goldin faced, Mark Albrecht commented, "He was the right person for the job." Prior to his joining, "The organization [was seen as] 'the gang that couldn't shoot straight,' 'Where's NASA going?' 'Lost in space,' 'Loss of focus, loss of performance.' And today it's seen as a bright, shining example of the kind of technology advancement, purposeful, positive element of U.S. foreign policy, U.S. civil research and development program . . . it's the kind of vital program and agency that I grew up with in the sixties when people had attributed it to Apollo and the Cold War . . . Dan and George deserve enormous credit."

Leadership Insights

- Leaders develop a network of trusted advisers, depend on them and are willing to listen their recommendations.
- Trust the advice of technical experts.
- The correct decisions are not always popular decisions.
- To find the path forward, understand how you got to where you are, visualize where you want to go then work backwards to where you are today.

CHAPTER 14

A Weekend in Reston

"That weekend saved human
spaceflight for the 21st century."
— GENERAL TOM STAFFORD, COMMANDER, ASTP

The International Space Station is the brightest object in the night sky after the Moon. It orbits the Earth at five miles a second, a beacon of hope for the future of space exploration. It is the result of the collaborative efforts of five major international partners, 16 countries, and is one of the most ambitious collaborative efforts undertaken in history. In June 1993, it was one vote away from being canceled by Congress.

When Goldin joined the agency, he immediately recognized the problem with the space station: "That was a very serious problem when I got to NASA. The Space Station Freedom had spent its entire budget, and when I wanted to see what there was to show for it . . . there wasn't anything built."[1] The administration had little patience with NASA and Goldin understood his longevity would be determined by his ability to reduce the cost of building the space station. Without a solution the program would be canceled, and the future of human spaceflight would be in jeopardy. Goldin had to act quickly.

Leadership

> "The best leader is the one who has sense enough to pick good men to do what he wants done, and the self-restraint to keep from meddling with them while they do it." — THEODORE ROOSEVELT

The *LA Times* writer Gregg Easterbrook commented that "Goldin is unique in NASA history in several respects. He's the first space administrator not in awe of NASA culture — though TRW, his old employer, is an aerospace contractor, it is one that historically has had hostile relations with NASA. He's the first NASA chief without a career stake in the manned-flight program, which NASA has long used as an excuse to avoid rational thinking about space budget priorities."[2] While Goldin initially focused on changing the agency, Abbey was still working closely with Albrecht on the Space Council, and the two provided Goldin with a number of options to consider as he transitioned his leadership team. Abbey enjoyed working with Goldin, describing the new NASA administrator as "sharp and intelligent — a brilliant and visionary engineer."[3] The recommendations helped.

Normally a new presidential administration would change the NASA administrator, but Bill Clinton decided to keep Goldin in his role. After the transition *Washington Post* reporter Kathy Sawyer wrote, "With the civilian space program braced for a historic shake-up, the White House intends to retain NASA Administrator Daniel S. Goldin, a Bush holdover, as shaker in chief, according to a White House source."[4]

Goldin had never been a fan of Space Station Freedom. Shortly after his appointment he met with Aaron Cohen, acting deputy administrator; Marty Kress, associate administrator for legislative affairs;

and George Abbey to discuss options. Goldin wanted to redirect the program, but Kress and Cohen argued that would lead Congress to cancel the program. Little progress was made and after the confirmation of President Clinton, Goldin was asked to come to the White House on Friday afternoon in the first week in February to meet with science adviser John Gibbons. Gibbons got to the point. "We're going to cancel Freedom."[5] The new Clinton administration's focus on the budget had quickly targeted the over-budget program and concluded it was going nowhere. Goldin asked for some time to respond but the new administration was in the process of preparing a new budget for Congress and time was not an option. "You have the weekend,"[6] Gibbons responded. Goldin returned to NASA headquarters wondering how he would save the future of human spaceflight.

Proposal

> "George Abbey personally saved the space station."
> — TOM STAFFORD

He had 48 hours to find a solution. The first thing he did was call George Abbey. Abbey in turn got on the phone and called Tom Stafford. The relationship they had formed over the years in Apollo, ASTP and shuttle would prove to be a critical element in finding a solution. Unfortunately, Stafford was in Florida. He couldn't return but offered the use of his office in Alexandria and called his secretary to have the keys delivered to Abbey. Abbey called Joe Shea, asking him to come for the weekend. Shea was one of the most brilliant engineers Abbey had ever known, who had left NASA in the aftermath of the Apollo 1 fire. Also included in the group were renowned spacecraft designer Max Faget, Mike Mott from the

Synthesis Group and John Young, the brilliant engineer and commander of the first shuttle flight who had also participated in two Gemini flights and two Apollo missions. Mike Mott would go on to become an integral part of Goldin's team as the associate deputy administrator where he functioned as a chief of staff, similar to a chief operating officer, "to work on the day-to-day activities across all of the enterprises of NASA and in all of the functional areas."[7] With pencils and pads of paper belying the high-tech project they were trying to save, they got down to work around the conference table. Together they were visionary realists, each an expert in their own right, crafting a plan that would create an orbiting platform to prepare for human spaceflight in the 21st century.

Goldin outlined the challenge: a program that had to fit within a $2 billion a year envelope or else there would be no station and no future for human spaceflight. The group included a brilliant visionary, the penultimate influencer, two brilliant aerospace engineers and the most experienced astronaut in history. George Low had tragically died of cancer in 1984 when he was 58, but through his close working relationship with those who were there, his perspectives were well represented. The future of human spaceflight was in good hands.

It took little time for Faget to outline his vision of a modular space station that could be operated autonomously during construction. Pressurized crew modules could be added when ready with the additional modules linked together to form the final configuration. Both George Low and George Abbey had been supporters of the idea years before, but it had not gained traction in the private sector when Faget left NASA to work on a commercial project he called the Industrial Space Facility. The Russian Mir space station had successfully utilized a similar modular concept to expand its size and capability in its first six years of operation.

The proposal was unanimously supported by the group. Joe Shea advised, "You need to take a hard look at simplifying the subsystems and streamlining the management of the program." Faget added a critical caveat, "You also need to go back to the approach that was the key to our past success and manage the program from Houston."[8] It was a clear reminder that the direction for the future can be determined by the course of the past.

ISS

The success of that weekend was measured by Gibbons's reaction to the proposal. Goldin met with Gibbons at the White House and explained the plan by laying out matches on the table, as Faget had done on the weekend. Gibbons responded, "If you can do it for $2 billion a year, it will work."[9] Gibbons was true to his word, the renamed International Space Station Project was approved, and it would be led from Johnson Space Center. Within four years the first ISS element was launched and within seven years of the critical weekend get-together, the first astronauts and cosmonauts would visit the station, beginning what has become a continuous human presence in space spanning two decades.

How did a one-weekend meeting of six people save the space station project when a team of thousands working on Space Station Freedom for the better part of a decade had been unable to meet cost and design requirements? In part, success was due to the choice of who was there and who wasn't. Dan Goldin's willingness to listen to the assembled experts was another, but success would ultimately come from adhering to the guidelines for large space projects outlined as legacies in the Synthesis Group report.[10] With the exception of Goldin, each participant in the weekend meeting had either been a part of, or an adviser to the group. Abbey's unique ability to understand what needed to be done and his commitment to doing the right thing had been pivotal in developing the recommendations of the Synthesis Group, in the creation and operation of George Low's Apollo configuration control board, and the success of the two-day redesign and rescue of the space station program. Some would call these legacies aphorisms for future spaceflight. Each of the guidelines and pitfalls were based on the collective wisdom of some of the best aerospace engineers and leaders in the history of human spaceflight. They stand as reminders, collected wisdom from those who sent humans to land on the Moon and return successfully to the Earth to the next generation of leaders. Reading them is easy, embracing them can be harder. **Sometimes doing the right thing is not always expedient or popular.**

Leadership

The team Abbey brought together knew that the right way to build a space station would be "based on simple interfaces between subsystems and modules, would make maximum use of modularity . . . to build on the capabilities established by prior [missions]. To have redundant primary and backup systems. Design in redundancy versus heavy reliance on onboard maintenance."[11] Each of those guidelines that had proved critical to the success of the previous era of human spaceflight would be critical elements of the success of the International Space Station program. The NASA culture, centered on complex, high-stakes technical projects that would repeatedly make the seemingly impossible possible, had learned the value of having "clean lines of management and responsibility for all elements of the program, realistic program milestones that provide clear entry and exit criteria for the decision process . . . [a process to] hire good people, then trust them."[12]

After the Apollo 1 fire Gene Kranz challenged his team to be tough and competent. At NASA, competency builds trust that becomes the basis for organizational resilience — a factor that was critical in returning to flight after the losses of Apollo 1, Challenger and Columbia. There are many lessons from human spaceflight for leaders aspiring to create peak-performing teams. **Hire the best** people you can find, create an environment where they continue to learn, listen to them, trust them and give them the tools they need to succeed. Trust and verify — George Low and George Abbey learned the importance of holding teams accountable and mastered the art of operationally insightful questions that helped teams achieve success. History will never know what would have happened without that approach. It is clear that two of the most important accomplishments in history, humans landing on the Moon and an international research platform in space, would not have happened without that approach. It is a profound lesson for all leaders.

"George Abbey personally saved the space station," said Apollo astronaut and longtime NASA adviser Tom Stafford. Stafford credits Abbey with bringing Apollo-style management practices to the station project, which had been mired in bureaucracy and engineering missteps. "If he'd gotten run over by a truck . . ." Stafford said, "we still wouldn't have anything up there."[13] While that statement is true, George Abbey would be the first to say that it was the thousands of dedicated civil servants, contractors, cosmonauts and astronauts that led to the success of the International Space Station program.

Leadership Insights

- How we do things determines how safe we are. Safety is not a simple act, it is a corporate habit.
- Failing to keep, read and share decision diaries, corporate lessons and debriefs sets the stage for future failures. Learn from history to avoid repeating what has not worked in the past.
- Competency builds trust and becomes the basis for organizational resilience.

CHAPTER 15

Working Together

*"We are at a point in history where a proper
attention to space . . . may be absolutely crucial
in bringing the world together."*
— MARGARET MEAD

"This was the major foreign policy initiative of President Reagan,"[1]
commented former Canadian Space Agency president Mac Evans
on the NASA plan to build Space Station Freedom. Peggy Finarelli
of NASA's international office recalled it was "a leadership issue
very much in the context of the Cold War. We were challenging the
Soviets in the high ground of space. We had to say that Freedom
would be bigger and better than the Soviet space station."[2] Evans
recalled it was to be used "as a foreign policy initiative to show the
rest of the world that our way of life was better. NASA just happened to have the right idea, at the right time; [a] space station and
an administration [that] grabbed onto it."[3]

The plan for Space Station Freedom was built on a collaborative
international partnership. Evans noted, "the NASA administrator
Jim Beggs came around . . . to Canada, to Europe and he went
to Japan and offered [the] invitation."[4] NASA and its potential
partners had started work on defining the space station and what
each partner might contribute. This helped each of the partners to

prepare for the intergovernmental agreements and MOUs (memorandums of understanding) that were signed between each agency and NASA in September 1988.

Based on the successful Apollo-Soyuz mission, many within NASA recognized the collaborative opportunities to work with the Russian space program, but Russia was not invited to participate. There were a number of other factors that affected the U.S. relationship with the Soviet Union to work together in space, including the Strategic Defense Initiative also announced in Reagan's 1984 Presidential address. Senator "Spark" Matsunaga from Hawaii warned of the dangers of weaponizing space and emphasized the importance of working collaboratively with an "internationally developed space station as an alternative."[5]

The tragic loss of Challenger in 1986 affected all the spacefaring nations and the world reacted in sorrow. The European Parliament held a moment of silence for the crew while calls and cables expressing sympathy poured into the White House. At the United Nations, the Secretary General expressed his "profound sadness" to Reagan: "Truly the entire world will grieve this tragic loss of life incurred in the advancement of the frontiers of human knowledge."[6] Despite the risks, each of the partners in the space station program wanted to continue. Reagan expressed what many felt: "The future doesn't belong to the faint-hearted. It belongs to the brave. The Challenger crew was pulling us into the future, and we'll continue to follow." Space Station Freedom was the plan NASA wanted to follow.[7]

Options

As the decade came to a close, the budgetary challenges of the Freedom program were watched closely by the international partners. Goldin felt "Space Station Freedom was being built as a gargantuan job to show the world we could build a bigger and better Space Station.

No. . . . We want to show the world we could work together. And it doesn't have to be that expensive."[8] The partners were confident that the international agreement would protect the project and their investment in the technology they had developed, yet the future of the space station was far from assured.

The Reston replanning weekend bought NASA time for an external review to assess the new options for the space station. Recognizing the strategic importance of an external review, Goldin immediately contacted Chuck Vest, president of the Massachusetts Institute of Technology (MIT), who he thought would be "fabulous" for the role. Vest eagerly accepted the task, commenting, "The panel has only one goal — to provide the federal government with an accurate assessment of the various options proposed by NASA's redesign team. This will include determining whether design objectives are met and providing critiques of the proposed management methodologies and cost projections."[9] The international partners were asked to participate in the review and when finished the panel concluded, "NASA has met that challenge, offering a plan that will substantially reduce costs to taxpayers, improve management, preserve research, and allow the United States to continue to work with its international partners and keep its international commitments."[10]

Mac Evans felt the foreign policy initiative was an important element in securing the future of the project: "The message was sent through the diplomatic channels that this would be a significant disaster [if canceled]. The Europeans played a major role and Canada played a significant role . . . this is an international treaty and [if] you're going to pull out . . . [it] is going to be a huge foreign policy debacle."[11] This was clear to the Space Council as well. Mark Albrecht recalls the change in approach to "how we might exploit international cooperation, not simply as another fragile element on a coalition built to support the space station, but really a fundamental and integral foundation on which to build. . . . That, of course, led the way when NASA undertook the full redesign of the space station."[12]

> "We make progress when there's very strong political support for the program." — MAC EVANS

The Clinton administration welcomed the findings of the Vest panel and President Clinton called for "the U.S. to work with our international partners to develop a reduced-cost, scaled-down version of the original Space Station Freedom. At the same time, I will also seek to enhance and expand the opportunities for international participation in the space station project so that the space station can serve as a model of nations coming together in peaceful cooperation."[13] Some felt that Dan Goldin did not have the political savvy to effectively maneuver the tough Washington environment, yet within two years of his appointment he succeeded in restructuring the agency and managing other programmatic issues while rebuilding support for the space station program.

Red and Blue Teams

Changing organizations is challenging at the best of times, yet Goldin (described by some as a "Washington outsider") succeeded where many thought he might fail, and others hoped he would fail. There were a number of key factors that played a role in his success. He had a clear vision of what he wanted the agency to achieve, he understood NASA's history and the administration's goals and he had a plan to turn his vision into reality.

Shortly after starting, he created two headquarters teams, a red team and a blue team, each tasked with recommending options for improving agency efficiency. The different programs were reviewed by both teams, with one serving as a critic of the other. Goldin had developed this concept when he was at TRW: "The blue team proposes, and the red team beats them up. . . . This was a new concept that wasn't done at NASA."[14] This approach, while new to the NASA headquarters team at that time, underscored **the value of disagreement, discourse and debate to determine the best course of action**. This approach had been a critical element of success in the Mercury, Gemini and Apollo programs and was part of the early culture at NASA.

Engaging staff in discussions about organizational change can play an important role in getting support for new ideas and new ways of doing things. Goldin was pushing hard, trying to get the teams to develop strategies to streamline NASA programs by as much as 30 percent. This would be a challenge for any organization but getting buy-in from all or part of the teams helped Goldin gain internal support for change. In a resource-constrained organization, with so many competing technical programs, there would never be widespread support for change initiatives. First followers or early adopters of the plan are critical, and through the work of his red and blue teams he was able to get a degree of internal support for the changes he was making. As a leader, Goldin relied on the input of a relatively small group of technical experts who soon became trusted advisers. The decisions he made were strengthened by the recommendations of internal or external expert advisory teams when necessary. In the world of high-stakes leadership, it was a successful strategy.

World Stage

There were international political changes happening in the early nineties that played a significant role in the future of the new vision for the International Space Station. As the Berlin Wall fell and the Soviet Union became the Russian Federation, many "began to see unique opportunities for engaging and involving the former Soviet Union in the space station program. We saw the opportunities [to] reduce the cost of the space station and were interested in how to move the schedule of space station forward, because we wanted to get on beyond the space station on to further space exploration."[15] The Russian space program had significant expertise in heavy lift capability, advanced rocket engine technologies and long-duration human spaceflight experience on the Mir space station. Despite the many compelling operational advantages associated with inviting the Russian space program to join the international partners, many at NASA did not support the idea.

Incorporating redundant systems is one of the primary strategies in controlling the risk of human spaceflight. Inviting the Russian Space Agency to join the International Space Station partnership provided additional launch and landing capabilities as well as the operational experience of the Russian mission control team. Goldin recalled, "Here was the problem, America had the shuttle. We had one vehicle to go build it and operate it. And there was no way that Space Station Freedom would have survived with just the shuttle. We needed functional and numerical redundancy and access. So, the beauty of international cooperation is we had the Russians, having a capacity for launching supplies to the station and people."[16]

Redundancies give systems and organizations the ability to withstand failures. Through the efforts of George Low, Chris Kraft and others, NASA had developed a fail-ops, fail-safe approach to controlling risk. The first failure may have mission impact, but the system remains operational. After two failures the system is not fully operational, but it remains safe. It takes three failures for safety to be compromised, leading to activation of contingency plans or abort scenarios. With the focus on foreign policy and cost reduction, few imagined that in the future NASA would have to rely solely on the redundant launch and landing capacity provided by the new Russian partners. The small group that understood the benefit the new partner would bring to the space station included George Abbey, Tom Stafford, John Young, Chris Kraft and Dan Goldin, the weekend warriors who replanned the program.

Respect

Effective collaboration starts with respect. Collaboration can't be forced; there must be mutual recognition that each organization will benefit. There are many reasons that different organizations or teams want to collaborate. In the case of the space station, it was a unique blend of foreign policy, cost reduction and benefiting from operational capacity and expertise. As events unfolded in 1994, NASA and Russia began to work together during the Phase 1 NASA-Mir program, developing a mutual trust and respect that would form the basis of a strong alliance among all the international space station partners.

The new program manager Bill Gerstenmaier immediately appreciated the importance of understanding, respecting and trusting the people you're collaborating with when he was assigned as the program operations manager of the NASA-Mir program. In getting ready to support the first long-duration flight of Shannon Lucid, the NASA astronaut who was about to work on the Mir space station, he quickly realized that going to Russia for a couple of weeks to prepare for the mission would not work. "I was supposed to go for two weeks, and then come back for two weeks and then go for two weeks and come back for two weeks. Well, I went, and then I figured out that if I'm going to really support Shannon on orbit, I have to stay continuously. I told my wife . . . I think I'm here, like, for a long time. I ended up being there almost a full year. But I think what happened during that time was the Russians had not seen Americans come over very often. . . . Then they saw me come and stay, which was unique." With extensive long-duration spaceflight experience, the Russian team understood what it meant for Gerstenmaier to be away from his family. "They kind of adopted me. They also figured out I could actually do stuff, I could read rendezvous procedures, I could understand systems on Mir. . . . They knew that I wasn't just the typical person to visit. . . . I actually could add value to them. Then they kind of adopted me into the system, and I became part of their mission control team."[17]

Partnership

Not surprisingly, astronaut Mike Foale had taken the same approach when he was assigned to a long-duration flight on Mir. He and his family moved to Russia, learned to speak Russian and integrated into the Star City community. Like Gerstenmaier, through his competency and commitment to embracing the Russian spaceflight culture, he built respect with and gained the trust of the mission control team. That trust was critical in dealing with events in space; it was also an important element in the trust NASA senior management had in the Russian Space Agency.

Collaboration was not always easy. The partnership would face challenges, with the combined repercussion of an eventual fire, collision and depressurization of the Mir Space Station that occurred in later missions. These events led to a loss of congressional confidence in the NASA-Mir program and increased doubts about the value of the partnership: "Members of Congress and others yesterday charged that the risk to American astronauts has reached the point that it overwhelms any benefits to be derived from their work aboard Mir. Rep. F. James Sensenbrenner Jr. (R-WI), chairman of the House Science Committee, yesterday asked NASA Administrator Daniel S. Goldin to conduct a 'top-to-bottom' evaluation of the safety of Mir."[18] "Otherwise," added the congressman, "no American astronauts should be sent to Mir."

Goldin had always been focused on crew safety. "Believe me, I don't sleep nights. And there's only one thing that has been on my mind for weeks now — the safety of our American astronauts."[19] Both NASA and the Russian Space Agency commissioned a number of expert teams to review what had happened and recommend changes and options to continue the program. It was a true test of the partnership.

In his testimony as the chairman of the Congressional Mir Safety Hearing, James Sensenbrenner, Jr., remarked, "There has been sufficient evidence put before this hearing to raise doubts about the safety of continued American long-term presence on the Mir." Drawing from the reports of the three NASA convened reviews, program manager Frank Culbertson responded by focusing on "two broad questions." He asked, "Is there sufficient value and benefit to be gained from continuing the missions aboard Mir?" And "Can we conduct those missions safely?" He addressed each of the questions about safety individually using data from the numerous investigations, and he returned to the question of value by saying, "There is much more to be learned. When you are exploring new territory or preparing yourself to take a major step into the unknown, who can say when you have learned enough?"[20]

Goldin committed to continuing the partnership leading to the success of the program and solidifying the relationship between NASA and the Russian space agency. "Phase 1 [of Shuttle-Mir] succeeded for three reasons. First," commented Culbertson, "it succeeded because we had the unwavering support and guidance of key leaders such as George Abbey, Dan Goldin, and Yuri Koptev despite the most intense political pressure from outside the two space agencies." It was effective leadership built on the respect and trust between the two programs that responded to the crises and grabbed success out of the arms of failure. Recognizing the importance of the partnership, NASA astronaut David Wolf commented, "It's easy to be good partners . . . with the Russians when things are going easy. But it's when things are difficult that we really can show what good partners we will be."[21]

When Evans spoke of Goldin's tenure as administrator, it is not a surprise that he referred to Goldin as "an inspirational leader — a true internationalist — a man of mission, passion, and conviction. He changed what humans do in space — and how they do it — forever."[22]

Leadership Insights

- Include recommendations of internal or external expert advisory teams when necessary.
- Incorporate redundant systems as a strategy to control risk.
- Effective collaboration starts with mutual trust and respect.

CHAPTER 16

Permanent Presence

"I think the International Space Station is a great place to live for a year."

— SUNITA (SUNNI) WILLIAMS, NASA ASTRONAUT

The beginning of this millennium will forever be remembered as the dawn of a new era of human spaceflight. There had been many remarkable achievements by NASA and the Russian space agency in the preceding 40 years, but the launch of Bill Shepard, Sergei Krikalev and Yuri Gidzenko November 2, 2000, was different. This was the threshold of an international collaborative effort to build and use one of the most complex technological projects in history. Since then, astronauts and cosmonauts from many nations have been living and working together in space, conducting research and demonstrating the international cooperation essential for sending humans on longer voyages, farther into space. It will be remembered by future generations as the point in history when humans became a spacefaring species.

The technological achievements alone were remarkable. The large modules and supporting equipment were delivered on 42 assembly flights, 37 on the space shuttle and five on the Russian Proton/Soyuz rockets. It is 357 feet wide, 167 feet long, has approximately 2.3

million lines of computer code with 1.5 million lines of flight software code running on 44 computers communicating via 100 data networks in the U.S. segment alone. The software uses 350,000 sensors to monitor the myriad onboard systems to ensure station as well as crew safety and health. The growing list of visitors includes more than 240 individuals from 19 countries as of this writing, with 15 nations working together supporting its operation and management. At a cost in excess of $100 billion, the ISS is the most expensive collaborative project in history, and is recognized by most as a unique laboratory orbiting the Earth. It was replanned one weekend in Reston, Virginia, one vote away from cancellation during its approval, and was built in roughly 13 years. It is as much a story of leadership and teamwork as it is a technological success. Once again, George Abbey was at the center of the story.

The Vest Committee's recommendation was supported by President Bill Clinton and narrowly survived a congressional vote, gaining approval by a one-vote margin in June 1993. There were many other foundational elements necessary to build what was then known as Space Station Alpha. The NASA-Mir program was one of those. At the end of June, a delegation including Goldin, Abbey and the new head of the Space Council, Bryan O'Connor, went to Moscow to begin discussions that would ultimately lead to a $400-million agreement for 10 shuttle flights to the Mir Space Station. Later that year, the Clinton administration formally invited the Russians to participate in the International Space Station program.

Contractors

Realigning the activities of the four major contractors of the Space Station Freedom project into a single prime contract was another critical element that had to be resolved. Goldin met with the CEOs of Boeing, McDonnell Douglas, Rockwell and Grumman to present them with the new plan. Fundamentally it was a case of saving a government initiative that was in jeopardy, and finding a new way for the contractors and NASA to work together or the contracts and the space station would be lost. Goldin advised the CEOs that NASA would select a prime contractor without a formal competition and the remaining contractors would be overseen by the new prime contractor. NASA announced Boeing as the prime contractor in August 1993, and six days later Goldin submitted a determination to Congress stating, "It was in the public interest to use other than full and open competition to make Boeing the single prime contractor for the space station."[1] Boeing hoped to take advantage of the expertise of McDonnell Douglas, the Rockwell Rocketdyne team and Grumman; Boeing spokesperson David Suffia commented, "Our role is to lead this partnership of all the contractors. NASA is still the customer."[2] It was also announced that Johnson Space Center would lead the redesign, and Goldin wanted Abbey back in Houston.

High-stakes leadership is built on the foundation of trust and competency. Goldin, like Abbey, was committed to finding the best people and entrusting them to lead key NASA projects. When Aaron Cohen, director of Johnson Space Center, announced his retirement in August 1983, Goldin selected Carolyn Huntoon for the role. Formerly the director of the Space and Life Sciences Directorate at Johnson Space Center, she was working as an acting associate director for Cohen. She recalls speaking with Goldin "on two or three different occasions about the possibility . . . when he brought it up to me the first time, I told him I was reticent to take the job because it was mostly engineers there at the center. He thought my background, my education, my talents for getting along with people . . . was what the center needed at that time. . . . [When] he came back to me the second time, I [had] talked with several other people . . . and they all encouraged me . . . to do it, because it was a remarkable job. And I'm glad I did. I had a great time."[3]

Assessment

Her science background and experience in the Space and Life Sciences Directorate had reinforced for Huntoon the importance of getting "at least two or three" different versions of a story before making a decision and where possible getting the best information to make data-driven decisions. One of her first initiatives was a major reorganization of the Johnson Space Center to streamline it around the shuttle and station programs. "It was a period of assessment for the Center, of rebuilding some of the organizations that needed the attention and focusing on what the change in the agency was going to be with the shuttle and the station being the primary focus,"[4] she said. Part of that reorganization included Abbey as the new deputy center director.

> "What's important is that you have a faith in people, that they're basically good and smart, and if you give them tools, they'll do wonderful things with them." — STEVE JOBS

Abbey admired Huntoon and had recommended her to Goldin as a candidate to become the next center director, but he was reluctant to take on the new role of deputy center director as he still felt he was needed in Washington. He expressed his reluctance to Goldin. Goldin called Stafford, who was in New Zealand at the time, for assistance. Stafford in turn called Abbey, to convince him to take the position. "You're the only one who can make the ISS work."[5] Abbey reluctantly agreed and convinced Mike Mott to take his previous position as the associate deputy administrator. Mott recalled, "I thought we had all the difficulties that you can imagine. We had the obvious technical challenges. . . . I thought we had a number of political issues that were going to be difficult. We were going to involve a lot of different NASA [centers] and some other roles as well. We were going to involve all sorts of agencies within the federal government. . . . I really felt like that was going to be the challenge of really getting through, and I thought we were going to have an internal NASA challenge"[6] adding the new shuttle flights to Mir to the existing manifest. Johnson Space Center would quickly become the epicenter of these many challenges and Abbey returned to Houston as the deputy director, committed to doing what it took to succeed.

Recruiting

It took, time, teamwork, competency and commitment. Building the teams to support both the shuttle and station programs meant recruiting the best and the brightest engineers. "We had to recruit people. One of my goals was to recruit some younger people into the Center with some qualifications, because we were losing people. People were retiring. That was the period of time when people who had been at their heyday . . . during Apollo were now retiring from the Center and leaving some voids," Huntoon recalled. She built a network with universities to "reach out and get the best people in the country thinking about our problems" and leveraged relationships with the private sector when dealing with new technologies.

She and Abbey quickly began to utilize each other's strengths. Huntoon focused on running the Center and building the scientific capacity and networks that would ultimately use the space station as a research platform. Abbey took on the operational challenges of running the shuttle programs while building the infrastructure and capacity that would be needed for space station construction and utilization. At all levels it was a team effort. Despite the many challenges ahead, the change in pace energized the center. "I think talking in the 'I' part is hard for me, but it's also hard for anyone that's worked in an organization like NASA, because it isn't an 'I.' It's always a 'we.' There's always a lot of people involved, [it's] a team effort,"[7] Huntoon said, reflecting on the role she played.

While Huntoon's networking efforts were focused externally, Abbey focused on building his internal network as he had done working with George Low in the Apollo program. In those days both Low and Abbey would frequently drop by mission control, often in the middle of the night, to chat with the flight director and others about the issues and challenges they were dealing with. It is hard as a leader in large organizations to stay connected with the concerns of front-line team members. Low had visited, asked questions of and listened to the thoughts of the NASA team. He also infamously visited, observed and chatted with the Apollo contractor workforce to see how things were done in those organizations. When he candidly presented his conclusions to the CEOs of those companies, change rapidly ensued. Emulating his mentor, Abbey sought feedback on the real issues and concerns on the many changes that were being implemented. He cared for his people. Space shuttle commander and head of the Flight Crew Operations Directorate (FCOD) Jim Wetherbee said, "He worked both on the technical side and the social side. He doesn't get a lot of credit for working on the social side. . . . But I was working very closely with him watching how much he did care for the people who were doing the job."[8]

Focus

Abbey realized the operational pressures of flying the shuttle missions while concurrently building the space station added risk to the organization. Controlling that risk meant a relentless commitment to operating with the highest levels of excellence, quality and safety. Through his participation in flight readiness reviews, mission management team meetings, and visits with front-line team members, he subtly influenced the culture. "Too many leaders . . . too many managers would tell people that our mission is to fly safely, and what most of them assume that meant [is] the mission was to accomplish flights. George changed that message; he didn't explicitly say it, but he gave people who were listening to him the very clear understanding . . . that success is not flying missions, success is flying missions correctly with the highest possible quality,"[9] Wetherbee said. It can be intimidating for front-line team members to speak with senior executives in any organization. Abbey's late-night visits helped him focus the workforce on safety and quality; they also helped him validate the information he was getting through the management chain of command.

He attended flight readiness review meetings and mission management team meetings to ask probing technical questions in part to change the culture to focus on safety but also to verify that all the flight issues had been resolved. Wetherbee recalled, **"He would never pick a launch date until he was confident that . . . all the problems that needed to, had been solved.** There was the opposite of schedule pressure. He did influence people to work hard . . . [and] he was satisfied . . . [when] the technical issues [were] answered, or [there was] a plan to address the technical issues." That approach enhanced the safety of the program, noted Wetherbee: "If you look at the flight rate over the entire history of NASA, he had the highest flight rate. . . . He didn't make decisions to push [to launch], instead [he'd] delay and fix a problem. . . . He was always fixing them before it became a major issue."[10]

Huntoon and Abbey also had to develop the infrastructure at Johnson Space Center to accommodate new training facilities and a mission control center for Space Station operations. A life-sized, high-fidelity space station simulator was added to the shuttle simulators in Building 9, creating what would become one of the primary training facilities for long-duration crew members. Abbey had his eye on an unused building McDonnell Douglas had built near Ellington

Field in Houston in the early 1990s for the engineering and design of various components for the Space Station Freedom program. The existing weightless environment training facility that used underwater training to prepare astronauts for spacewalks was too small to be used to train for the approaching "wall of EVA" ("EVA" or extravehicular activity, the official term for spacewalking) that would be required to build the space station.

A new facility was needed. The refitting began in April of 1995 and what became known as the "world's largest indoor pool" was ready for operations 18 months later. Measuring 202 feet long, 102 feet wide and 40 feet 6 inches deep, it contains 6.2 million gallons of water, enabling astronauts to train on a full-scale mock-up of the International Space Station. Abbey also acquired a Super Guppy aircraft from Airbus to transport space station modules to the Kennedy Space Center, where they would be installed into the space shuttle payload bay on the space station assembly flights. In many ways the environment was similar to that of the Apollo program in the late 1960s, with thousands of details to be managed by NASA and the contractors who were following a rigorous schedule all while controlling the risk of failure.

Change is the one constant in large organizations. Goldin moved Huntoon to Washington in August 1995 to help with the creation of a new research institute that would follow the experiments and investigations once Space Station utilization began. Around the same time, she received an offer to go "over to the White House to work in the Office of Science and Technology Policy, to do policy work that affected the science in the whole nation."[11] Always the optimist, Huntoon reflected that "there's times in your . . . career where things change, external factors change, your own priorities change, and each time that it [did it] just seemed like it was the right thing to go do, and I haven't regretted any of it. [I] Had a great career."[12] Abbey was promoted to be the acting center director.

Competence

Anticipating the multifaceted needs of the space station program, Dan Goldin had asked Randy Brinkley, mission director for the Hubble Space Telescope repair mission, to become the space station program manager. The earlier Hubble mission meant Brinkley had to coordinate the work between the scientific community and different NASA centers to ensure the successful completion of the servicing mission.

After a 25-year career with the U.S. Marine Corps and a couple of years with McDonnell Douglas, Brinkley was an outside addition to the NASA team, quickly taking on the Hubble mission after he joined in 1992. Goldin felt that Brinkley's "human skills were great . . . because of his human qualities no one felt threatened by him, he was the right person to oversee the Hubble [repair]."[13] Brinkley recalled, "I think Mr. Goldin, whether rightly or wrongly, perceived me to have been successful in terms of building a team and taking the people from Goddard . . . as well as those at JSC, and bringing them together as an integrated team with one set of objectives, priorities, with clear lines of communication, authority, responsibility, and accountability. That was important to him going forward — having someone that wasn't necessarily locked into 'This is the way NASA has always done things.'"[14]

Goldin and Abbey needed the same approach leading the space station program at Johnson Space Center, and Brinkley was excited about working with the team. "The effort was being led by a number of really great individuals: Chet Vaughan, Bill Shepherd, Jim Wetherbee, John Young, a number of really bright people. . . . This also had great political impetus at the White House because it was viewed strategically as a way that we could incorporate the Russians into the space station,"[15] he recalled. "NASA is a wonderful organization and we had wonderful people in both the repair of the Hubble, and gosh, that certainly is the case on the International

Space Station." The NASA culture that a few years earlier had been "lost in space" had been transformed back to the Apollo-era ethos of **building the best teams by hiring for competency, giving them the resources they need and trusting them to do their job.**

Abbey implemented additional changes while continuing the rapid pace of the shuttle and space station programs. STS-63 Space Shuttle Commander Jim Wetherbee took on a new role as the acting deputy center director, Sue Garman continued as assistant to the director and former astronaut John Young became the associate director overseeing technical issues. Abbey focused relentlessly on safety, incorporating it into his remarks at his welcoming ceremony in the Teague Auditorium, and he implemented a Safety Awareness Day to complement the annual August open house. His work ethic permeated the center, and once again the center's parking lots were busy on evenings and weekends with team members trying to meet the ever-increasing demands of the shuttle and station programs. The center was thriving with the combined operational tempo of the shuttle program, the collaboration with Russia on the NASA-Mir missions and the all-encompassing effort to build the space station.

Integration

Goldin was pleased with the new changes and named Abbey as the permanent center director in January 1996 while designating Johnson Space Center the lead center for the space shuttle and International Space Station programs. The announcement consolidated the leadership of both major programs, and the future of human spaceflight, to George Abbey and his team. It would prove to be a daunting challenge that would engage Boeing in its new role as prime contractor while putting the program, the contractors, and the development of the Russian elements as well as those from the other international partners on a schedule that would bring all of these elements together for the first time in low Earth orbit.

Reflecting on the magnitude of the challenge, Abbey commented, "I was given responsibility for the space station and the shuttle programs when I took over the Johnson Space Center and at that point, [I] had really great concerns about the space station because it was very fragmented. The people were not working together, and the contractors were certainly not working together. There were feelings on the part of the international partners that they were not party to what was really going on. . . . And of course, we had just brought the Russians in which we felt was an essential part of the program.

"I felt that if we were really going to build a team," he continued, "we had to have openness and bring everyone in and treat them as an equal partner, treat all the international partners fairly and honestly and bring them into all the decisions, and also bring the Russians in as a full partner and do it openly and bring them into all the activities that were going on. The other essential part was trying to bring together and integrate the American contractors. I was very fortunate to have Boeing help me in that regard because Boeing bought Rockwell and they bought McDonnell Douglas. Finally, I had one contractor, really responsible for all the elements of the station. . . . I got to work directly with Boeing and make them integrate the elements . . . make them really perform and build [a] schedule that was doable. . . . I think that openness was really an important part of it . . . getting the issues out on the table."[16] The trust and respect built during the NASA-Mir program helped NASA and Russia work through numerous issues as the two lead partners. Abbey's inclusive leadership, when combined with his desire to treat everyone as an equal partner, solidified what could have very easily persisted as a fractured team.

Using a model similar to the early days of NASA, Abbey selected trusted, proven leaders to manage the shuttle and station programs. Randy Brinkley continued as manager of the space station program, while Tommy Holloway took on managing the shuttle program. Holloway began his career with NASA in the 1960s, working on the Gemini and Apollo flights. He was a flight director in mission control for the early space shuttle flights, became chief of the office and was subsequently named assistant director for the Space Shuttle program. He was a natural fit to manage the shuttle program. Abbey had clearly learned from his mentor George Low, but he soon found NASA engineers and the contractors repeating mistakes that had been made in the past. "You need to learn the lessons from the past. History doesn't repeat itself; [if they don't learn] people repeat history."[17]

Forum

It was time to take another lesson from the past and implement a configuration control board. Abbey suggested this to George Low as a method of managing the myriad issues associated with getting the Apollo program back on track after the tragic Apollo 1 fire. Drawing from the success of that experience, Abbey implemented a series of weekly meetings that were held every Saturday at 8 a.m. to ensure the station program was proceeding as planned. There was no fixed end to the meetings: they never lasted less than four hours and they would end when all of the work of the week had been discussed, action items noted, and a plan developed for the next week's activities.

There are very few leaders with sufficient influence to have the contractors in different parts of the U.S., personnel from various NASA centers, and partners from Canada, Europe, Japan and Russia, get together for a four- to six-hour meeting every Saturday morning. That influence came from mutual respect. Abbey respected and sought to involve all of the participants in the program in the process. The partners respected Abbey's leadership and always participated. He knew that the only way the program would succeed was for everyone to have a voice and a forum to be heard. The George Abbey Saturday Reviews, or GASSERs as they were affectionately called, were that forum. No one liked them, but no one missed a meeting. If planned work had not been done or problems existed, the meetings would run on until a new plan was developed. No one wanted to be the cause of a prolonged meeting and although it was never mentioned, the peer pressure to stay on track was undoubtedly a driving factor in keeping the program on schedule.

Abbey later reflected that this was the greatest leadership challenge in his career at NASA. "I had always thought I'd be responsible for a program where people work together, but there was no team working together. Teams are the elements of the contract and while the contractors were arguing with each other, the NASA team members did not feel they could work with the contractors. The contractors didn't feel like they could work with NASA. The partners didn't like the way NASA was doing things. And the Russians were just coming in to be a part of the program. The space station program, of all the activities I ended up being involved in, was probably the most difficult challenge that I took on. . . . In the space station [program], I was the one making the decisions on if there was going to be a program, I [now] suddenly had to make it happen."[18] He, the NASA team, the contractors and the international partners made it happen. Within two years the first elements of the space station were launched and within four years the first crew began what would become the longest uninterrupted human presence in space.

Leadership Insights

- High-stakes leadership is built on the foundation of trust and competency.
- Get feedback on the critical issues and concerns before changes are implemented.
- Fix small problems before they become big problems.
- Hire for competency, train for proficiency, give teams the resources they need and trust them to do their job.

CHAPTER 17

When the Smoke Clears

"Time was no longer like a river running,
but a deep still pool."

—— RICHARD E. BYRD

"The smoke was the most amazing thing," commented astronaut Jerry Linenger when describing the fire onboard the Mir Space Station. "I did not expect smoke to spread so quickly. It was a magnitude — I don't know, maybe ten times faster than I would expect a fire to spread on a space station. The smoke was immediate, it was dense . . . I could see basically the five fingers on my hand. I could see a shadowy figure of the person in front of me who I was trying to monitor to make sure that he was doing okay. But I really couldn't make him out, and where he was standing, he couldn't see the hands in front of his face."[1]

There are a number of potentially catastrophic situations that can arise on any spaceflight — fire and depressurization are two of them. Both happened on the Mir space station during the Phase 1 NASA-Mir program. Launching with Space Shuttle STS-81 crew on January 12, 1997, Linenger joined Mir-22 cosmonauts Valery Korzun and Aleksandr Kaleri who were close to finishing their 196-day mission. The two new cosmonauts, Vasili Tsibliyev and

Aleksandr "Sasha" Lazutkin, arrived on Mir February 10 with German researcher Reinhold Ewald, who would stay for 20 days before returning with Korzun and Kaleri.

From the moment they arrived the two new cosmonauts hit it off with Linenger. It looked like they were off to a great start of their long duration mission and Linenger looked forward to an undocking and fly-around in a Soyuz spacecraft, and a spacewalk that would make him the first American to conduct an EVA from a foreign space station. These two accomplishments would be overshadowed, however, by a fire that took place the evening of February 24.

With six crew onboard they had to supplement the oxygen supply by using backup solid-fueled oxygen canisters called SFOGs. After their evening meal, Linenger floated from the Base Block to the Spektr module to do some work, while cosmonaut Aleksandr "Sasha" Lazutkin went into the Kvant-1 module to activate the Russian-developed Vika chemical device (SFOG). It is a thin-walled three-part structure that creates oxygen through a chemical reaction that was used to supplement the Elektron system when there were more than three crew members on board the Mir station.

Linenger didn't get much work done. Shortly after he started a master alarm went off and when he went to see what was happening, he bumped into Tsibliyev and saw smoke coming from the Kvant-1 module. To Linenger, the fire resembled a box full of fireworks sparklers, all burning at once. The flame shot out about two to three feet in length, with bright bits of molten metal "flying across and splattering on the other bulkhead." The canister provided the fire with both oxygen and fuel. "It had everything it needed," he commented in a subsequent interview. Working together with Lazutkin and Mir-22 Commander Valery Korzun, "we immediately started fighting that fire,"[2] he reported later from Mir. "You had to react to the situation, you had to keep your head about you, so I guess it was just a matter of survival. The whole time they were thinking, 'We need to get that fire out.'" The situation was critical. The fire blocked the only path between the crew and one of the two Soyuz vehicles, leaving a way to escape for only three of the six men onboard the station.

The crew immediately began putting on oxygen masks as the space station's atmosphere became unbreathable. "I did not inhale anything, and I don't think anyone else did because the thickness of the smoke told you that you could not breathe. So, everyone immediately went to the oxygen ventilators. They protected us from inhalation injury."

Korzun ordered the one accessible Soyuz escape vehicle to be readied for evacuation, turned to face the fire and began using the Mir fire extinguishers. "When I started spraying foam on the hot canister, the foam didn't stick and had little effect. So, I switched to water and started using that," he said later, describing his efforts. The water turned to steam, adding to the smoke. Linenger stayed with Korzun, passing him fresh fire extinguishers while monitoring his level of consciousness. To stabilize Korzun in the absence of gravity, Linenger wedged his own legs into Mir's connecting tunnel and held on to Korzun's legs. "At one point," Linenger said, "I floated in front of his face, but the smoke was too dense, even at six inches," making it difficult for Linenger to see how he was doing. He resorted to tugging at Korzun's legs to be sure he was still conscious. Korzun recalled later, "Jerry kept tugging my leg. 'Valery, how do you feel?'"

Linenger, also a physician, kept an eye on his other crewmates. Anticipating a possible evacuation, Aleksandr Kaleri calmly worked at a computer, printing out re-entry information for both of the Soyuz vehicles. The oxygen canister eventually burned itself out, but smoke remained everywhere, even "in the distant modules at the very end of the cones." According to Lazutkin, "We even thought someone had switched the lights out in Kvant. That's how black it was."

The fire was over, having destroyed the SFOG canister as well as the panel covering the device. Linenger continued to monitor the health of the crew. "We set up a station for any respiratory problems that might take place. We had all the emergency gear in place. I did exams on all the crew members immediately following the fire, and then for 24 and 48 hours after that. I looked at oxygen saturation in the blood, checked the lungs — all the normal things you would do post-fire. From my assessment, I [didn't] see where anyone had any serious [smoke] inhalation damage, and it was due to good action by the crew to get into the oxygen masks quickly."[3]

Toughness

Long-duration astronauts have to deal with a number of different challenges and the NASA-Mir program provided the seven NASA astronauts who stayed on Mir with many insights into the nature of long-duration spaceflight. It was clear that there were significant differences between short- and long-duration missions beyond the aphorism "A short-duration mission is like a sprint, a long-duration mission is like a marathon." In fact, some ISS astronauts describe their experience as a continuous sprint for the duration of the mission. Given the principles for a successful long-duration mission are not always found in short-duration mission experience, the training team sought to discover the lessons learned from previous long-duration spaceflights and terrestrial expeditions throughout the world as far back as the era of the first Arctic and Antarctic explorers. It was another example of the assertion that the lessons for the future are written in the past.

Spaceflight is a competency-based team sport. Gene Kranz's challenge to his team to become tough and competent had evolved to become part of the core values in the NASA culture. Toughness, measured by resilience and commitment, ensured that the teams in mission control would do whatever it took to work with the astronauts in space to achieve success. Individual and team technical competency was critical; skills were honed relentlessly in simulators where crew trained like they would fly so that one day they would be able to fly like they trained. The experiences of the NASA-Mir long-duration astronauts added a new dimension to the competencies needed for astronauts to succeed. Behavioral attributes, leadership, followership and team skills became as important, if not more so, as technical skills in determining mission success. In preparation for multicultural long-duration missions to the International Space Station, the Behavioral Health and Performance Group developed a resource and training guide for long-duration crews that included field experience training in different spaceflight analogs. Long-duration crew members would train together in these analogs, in the wilderness with the National Outdoor Leadership School or in the ocean on the Aquarius Undersea Research Habitat through a collaboration between NASA and the National Oceanic and Atmospheric Administration, the agency overseeing use of the undersea habitat.

Space is a harsh, extreme environment, intolerant of error, that relies on complex spacecraft systems and technology for success. The crew wake every morning in the same place, to work with the same people, with only the timeline to demarcate work, meals, exercise and personal time. Over the course of a mission often those demarcations become less evident. Despite close attention to daily planning, many factors can affect what happens in the day-to-day life of a long-duration crew member. Participation in meaningful tasks has been attributed to success in managing the monotony of long terrestrial or space expeditions. The Phase 1 NASA-Mir crews found that having enough meaningful work to fill up the day was critical to keeping a good attitude and maintaining productivity. "If the schedule is active and meaningful, many of the other issues of isolation and confinement can be avoided."[4]

Motivation

The experiences of the NASA astronauts once again reinforced the importance of Dan Pink's research on motivation and performance. **The common attributes of motivation based on the principles of autonomy, mastery and purpose** directly apply to long-duration spaceflight.

Astronauts have reported a sustained sense of purpose conducting the extensive scientific experiments onboard the Mir and International Space Stations, selected through a rigorous NASA peer review process. Training for spaceflight is challenging, technically demanding and often described as "drinking from a firehose"; it is difficult to master all aspects of spaceflight. In fact, many astronauts would say that the true sense of teamwork in spaceflight comes from leveraging the expertise of each astronaut so that together they can master the mission objectives. Finally, crew autonomy has been an important balance since the earliest days of spaceflight when astronauts were referred to as "Spam in a can." Through the experiences of Mercury, Gemini, Apollo, Skylab and shuttle, the relationship between the team in mission control and the astronauts in space had achieved **the delicate balance between oversight and autonomy** to achieve mission success. Clearly the attributes to motivate long-duration crews to succeed were as strong in space as they were throughout the history of exploration on Earth. As any marathoner knows, motivation is important, but it alone does not guarantee success.

Teamwork

The foundation of expeditionary behavior relies on developing individual technical and behavioral competencies while emphasizing self-care, leadership, followership and teamwork. The Mir fire is a good reminder that spaceflight is characterized by ongoing challenges and low-level stress with the constant threat of a major catastrophe. Despite the monotony, the crew can be challenged at any point by potentially life-threatening emergencies. While the ever-present, but relatively low, probability of danger can be a stress, many astronauts have reported that the daily psychosocial stressors are more common and can have a greater mission impact. Understanding how to work together as a team is a fundamental element of spaceflight and the organizational culture at NASA where teamwork typically involves individuals collectively seeing themselves as a member of a group. Success depends on the competencies they bring to the group as well as their interdependence and interaction in working as a team to achieve success. Teamwork means a willingness to monitor one another's performance, to provide and accept feedback, to effectively communicate and to back fellow team members during operations.

The Mir-23 crew displayed excellent teamwork while simultaneously fighting the fire, ensuring the health and safety of the crew and planning for a possible abort scenario. The unique combination of individual and team skills in each of the crew, developed through years of generic and mission-specific training, contributed to their success in controlling the fire and arguably would have ensured success if the crew had been forced to evacuate on both Soyuz spacecraft.

Commenting on the challenges they faced, Linenger said, "[There were] two sets of difficulties we've had. One is the human difficulty of dealing with those things, and the other one is the space station itself. We've overcome all the difficulties. The ultimate test is we're still alive and well, we're all here exploring the frontier. On the other hand, it takes a lot of work, it takes daily attention, and it takes a lot of work from smart people on the ground looking over our shoulders and giving some guidance along the way. But we were able to overcome about as much difficulty as you can imagine."[5]

Often it has been said, "There is no I in team." The Mir-23 crew confirmed the power of working together to overcome adversity.

During the onboard crew transition before Linenger returned home he spent roughly six hours briefing Mike Foale, the fifth NASA astronaut to visit Mir, on the challenges of a long-duration space-flight. Foale was looking forward to the long-duration mission to Mir, despite the fire and other problems during the NASA-4 increment. He expected hard work, some discomfort, many challenges, and was looking forward to becoming part of the Mir-23/24 crew. Little did he realize that the Spektr module he was living in would be hit by a Progress resupply vehicle that accidentally rammed the space station, causing depressurization of the module that would have resulted in evacuation were it not for the teamwork of the crew.

"On the Mir, it's very easy to lose each other . . . you don't know where the other people are. . . it's not that the Mir is such a big space, it's because it's such a cluttered space. You're basically winding your way through, effectively, tunnels to go to one part of the station to another, so that equipment just isolates you from other parts of the station," Foale commented after his mission. Equipment stowage and cable management can be a challenge on any spaceflight; on long-duration missions it can be an even greater challenge. Over the years a number of cables had been installed between the Mir modules traversing the hatches that were designed to isolate the modules in the event of an emergency. This would be a problem for the Mir-23 crew.

Shortly after arriving, Foale focused on becoming part of the team and building the trust of the mission control team. "In those days before the collision, I learned a little bit about the troubles they've had, especially with the fire. Vasily talked about it quite a bit, and Sasha. At some point Sasha ... actually took me to where the fire occurred and showed me what he was doing and how the fire happened, and he gave me a long hour's description of every-thing that happened during the fire. It was very amusing. We were enjoying it. It was a good story, with serious undertones. But he wasn't making a big deal out of this; he was telling me a story, because I wanted to learn. And other times Vasily would talk [tell another story] about the near miss of the Progress docking . . . it was . . . a very close call."[6]

During the initial NASA-Mir missions the NASA astronauts spent the majority of their time operating experiments to learn more about microgravity. Linenger's participation in fighting and his clinical advice in recovering from the fire demonstrated the benefit of building a team that could leverage the skills of each of the crew. Foale worked hard to build on the teamwork that Linenger had started. Realizing the importance of communication working in a high-stakes environment, Foale had committed to developing his Russian language skills. Unlike his predecessors he spoke in Russian to the flight controllers in mission control. Foale recalled, "But over . . . a month, I got to know the controllers quite well and they got to know my voice well. I lost my inhibitions to talk to them on the radio. That kind of integrated me into the overall Russian operation of the station."[7] Initially the conversations were difficult due to the technical terminology but over time he became more proficient. With that proficiency came the opportunity to build relationships with the mission control team.

Collision

Described by Vladimir Lobachyov, head of Mir mission control, as "the most serious [space] accident in recent times,"[8] the collision involved the Progress M-34 resupply spacecraft that was launched in April 1997. Two days after launch it used an automated docking system to dock successfully with the aft port of Mir's Kvant-1 module. However, in June a test of the manually controlled docking procedure was conducted in which the Progress was undocked from Mir and Commander Vasily Tsibliyev used the TORU manual control system to attempt redocking. During the maneuver, the Progress spacecraft collided with the Mir Spektr module, creating a leak in the module and damage to the solar array.

Foale recalls, "One of the early thoughts that went through my brain . . . [was], 'You know, I've been here six weeks and I think we're going to be going home right now.' I was actually kind of sad. I thought, 'Well . . . that's a shame. I won't finish this whole thing.' I had set out here to be four and a half months, and now it's going to get cut short. This is a real emergency. And we had all the danger of getting out of there, but it was crossing my mind, 'This is a shame. I've only been here six and a half weeks.' [Then] I thought, 'You'd better focus on getting this sealed off here.'"[9]

Foale and Lazutkin started removing the cables to close the hatches to Spektr to isolate the module. In describing the challenge, he said, "A node is built with six holes. It's like a dice with six faces. Each hole has a hatch. . . . The thing is, you have to get them out of the way in Space Station life, so they've been tied up, and they've been tied up pretty securely. The biggest hatch . . . was really tied up pretty severely. I mean, we wasted about a minute trying to untie that hatch. And the pressure's falling, pressure's falling, so it's getting pretty frantic. . . . I was thinking, 'Things are getting pretty tense now.'" After closing the hatches, Foale noted, "I knew at that point we had isolated the leak, because I could feel the hatch holding in and then the pressure stopped dropping in my ears."[10] They had succeeded with minutes to spare.

Despite their initial success, the situation was still critical. During the orbit after the collision, mission control noted the station was spinning at about one degree a second. The crew would have to control the station; there was a quick interchange between mission control and the crew when the ground team asked, "'Guys, what's the spin rate?' 'We don't know' [the crew responded]. 'We've got to know how fast it's spinning.'" Sasha immediately started trying to determine the spin rate and axis. With a background in physics and guidance systems Foale went to the window. "[I] put my thumb against the window, looked at the stars and was able to tell the ground what the spin rate was. I called it down." Foale knew that this was the first time he had made an operational call on the state of the Mir to the ground, but they had no other choice and accepted his data. Lazutkin confirmed, "Well, yes, yes, Mike's right." The mission control team took that information and fired the engines in a blind mode to stop the spin. It worked. They said, "Did it work?" Foale looked out the window again, looked to the stars and said, "Yes, it worked."

The crew had worked quickly as a team to save the station, yet it was Foale's recommendation to mission control that proved critical in stopping the spin.

> "I really count all that we are doing together, America and Russia, to be extremely valuable to future cooperation on the Earth." — MICHAEL FOALE

For Foale, "The collision demarcates everything for my flight, not because of the terribleness of the collision; because it changed the whole condition of the station and the environment in which we worked." It also reaffirmed that competency is critical in a high-stakes environment. Competency, enhanced by effective communication, is the foundation for building trust. **High levels of trust are characteristic of high-performing teams in which roles are understood, goals identified and team members communicate candidly, with integrity and respect, to discuss ongoing challenges and successes.** Teamwork involves a willingness to understand that success depends on how the team members interact with one another. Being prepared to back up team members during operations and having the willingness to listen and accept assistance from other group members are fundamental skills of high-trust teams.

Each team member brings different values, attitudes, motivation and effort to the team. If the values of the individual reflect those of the organization, the shared values enhance team success. The Mir-24 crew possessed the individual technical competencies and built on the trust that had developed over the course of the mission to recover from the collision. The sincerity of Foale's attempt to learn to speak Russian, to understand the Russian culture, to build relationships with his onboard crew as well as the team in mission control, speak to his integrity and commitment to the mission. **Great teams continually build trust to be ready to respond to those moments when critical actions and decisions will determine success.**

Communication

Miscommunication can instantly destroy trust. Effective, assertive communication is a skill that must be developed in individuals and valued by organizations. In both the Challenger and Columbia disasters, ineffective and passive styles of communication contributed to the tragedies.[11] In the case of the loss of the Challenger, the Rogers Commission found that "NASA eliminated the element of good judgment and common sense when it came to communicating safety concerns." Before the launch, Alan MacDonald, Thiokol director of the space shuttle solid rocket motor project at the time of the Challenger accident, recalled, "I want you to get the engineers together and have them make an assessment of their concerns of the cold temperatures on the sealing of those O-rings. And I would like them to make a recommendation . . . what is the lowest, safest temperature to launch that we can tolerate, and I want that decision and recommendation to be made by the vice president of engineering."[12] The launch decision was ultimately made on the basis of a management poll, and the engineering team's analysis of the data recommending canceling the launch was either not presented, not heard or misinterpreted, resulting in the catastrophic loss of the crew and vehicle.

After Challenger, NASA implemented a number of changes to create a strong safety culture, yet 17 years later the Columbia Accident Investigation Board concluded that deficiencies in communication were a contributing factor to that disaster. It is widely accepted that organizational culture is based on shared values, attitudes, goals and practices that characterize the way work gets done. Incorporating effective communication styles into those organizational values is an important element of sustaining organizational change. The time between the Apollo 1 fire and the loss of Challenger was 19 years. Between Challenger and Columbia, the gap was 17 years. The changes in leadership between Apollo, Challenger and Columbia may have resulted in a gradual shift in the organizational culture that diminished the value of assertive communication, discussion and disagreement as tools to address differing viewpoints while seeking to understand and control risk. Recognizing the critical importance of communicating risk, the Synthesis Group recommended that future leaders "ensure that the administration and the Congress clearly understand the technical and programmatic risks . . . of the space exploration initiative [human spaceflight]."[13] History has proven that to be sage advice.

Assertive communication involving speaking up and listening was a valued skill learned in the Mercury, Gemini and Apollo programs. Leaders understood the importance of asking technical teams insightful questions that would uncover latent risks, enabling the teams to address technical challenges and control risk. As George Abbey observed in the early days at NASA, it was possible to have heated discussions debating a topic during a meeting and still remain friends in a social context. **Discourse was an organizational value used to control risk and optimize success. Experience** has shown that successful outcomes come from decision-makers asking the right questions, listening to the recommendations of highly competent trusted individuals and making a decision based on the merits of the discussion. That approach worked in Apollo 13, the Mir fire and collision and countless different scenarios. It is a lesson for leaders in any organization to build competency, trust your team members and listen thoughtfully to their recommendations. Great leaders value discourse.

Leadership Insights

- Spaceflight, like many corporate endeavors, is a competency-based team sport.
- The common attributes of motivation are based on Dan Pink's principles of autonomy, mastery and purpose.
- Competency, enhanced by effective communication, creates trust.
- Great leaders value discourse.

CHAPTER 18

Ultimate Trust and Teamwork

"How's the view?"

— CHARLIE HOBAUGH

"Power switch to battery." The call from mission control marked the official start of the second extravehicular activity (EVA), NASA parlance for spacewalk, of STS-118 in 2007. "Copy, going to battery," Rick and I (Dave Williams) responded. Having left the role as director of space and life sciences, I was once again assigned to a mission as a Canadian Space Agency (CSA) astronaut to help build the space station. Crewmate Rick Mastracchio had led the first spacewalk two days earlier in which we had installed the large fifth starboard element onto the space station's truss. The goal of this spacewalk was to remove and replace one of the four gyroscopes, which had started to fail.

Most spacecraft use gyroscopes to stabilize their position and orientation; this task would be critical to the future of the space station. As in our first spacewalk, this involved the carefully choreographed movements of the station's robotic arm integrated with our tasks. Success would depend on our crew working together as a team, trusting each person to perform the task to the best of their ability. There is no greater demonstration of trust than going out the airlock hatch with another astronaut into the extreme, harsh vacuum of space. With the wry humor of operators, it has been said that if you have a problem during a spacewalk, you may have the rest of your life to solve the problem. This spacewalk would test the skills of our team.

It takes time to build peak-performing teams. That is particularly true for spacewalks. The focus is typically on the astronauts performing the spacewalk, yet the crew inside the spacecraft and the team in mission control are critical elements in the ultimate success of completing the objectives. Typically spacewalks last between five and eight hours, and are made up of one or more primary, sometimes critical tasks, other secondary tasks and what are called "get-ahead" tasks that the crew can complete if they are ahead on the timeline. With the supply limitations in the life support systems of the spacesuit, staying on the timeline is critical.

As a crew, we learn during training to get ahead and stay ahead of the timeline, buying precious time to deal with an unforeseen issue that could arise at any moment. It is well known in the world of spacewalking that despite all the thousands of hours of planning and preparation for a few hours of execution, once the crew go out the hatch, whatever happens has to be dealt with. Hopefully things go as planned but unforeseen events such as hardware failures, stuck bolts, lost tools, spacesuit emergencies and false alarms occur and must be managed.

Halfway through the third spacewalk of our mission, Rick would discover a hole in the outer layer of his spacesuit. Likely caused from contact with a sharp edge of metal created when orbital debris struck the space station, the not-yet-leaking hole in his glove resulted in early termination of the spacewalk. On the mission's fourth spacewalk, Clay Anderson and I were twenty minutes into our tasks when the fire alarm went off inside the space station. We knew the crew and mission control would deal with whatever the issue was; our job was to continue on the timeline until advised otherwise. It turned out to be a false alarm that had no mission impact as we had continued to focus on our tasks.

Training

The mantra "tough and competent" has been embraced by every spacewalking team in history. We survive based on a shared trust built on our individual competencies, effective communication and ability to deal with unforeseen circumstances. Success depends upon our individual and shared judgment, skills and knowledge. There are many different training environments used to help astronauts develop these skills. From Ed White's performing the first NASA spacewalk during the Gemini 4 mission in June of 1965 through Apollo, Skylab and into the shuttle era, astronauts have used underwater training as the primary tool to learn spacewalking skills. "Frictionless" air bearing platforms, parabolic flight, a partial gravity simulator called POGO and more recently, and perhaps more effectively, immersive virtual reality has been used as well, yet the majority of training takes place underwater. To prepare for the spacewalks needed to build the space station, NASA built the Neutral Buoyancy Laboratory (NBL), the world's largest indoor pool. Whatever the training environment, the crews learn about suit systems, spacewalking tools and techniques, how to perform different tasks and most importantly how to work together as a team.

All astronauts participate in generic spacewalk training. Mission-specific spacewalk training is frequent, focused and intensely scrutinized to ensure crews will be ready to perform the tasks in space. It is an opportunity to learn, to hone the skills of individuals and the team; it is also a place where failure is an opportunity to learn. The spacewalk training for STS-118 started shortly after our crew was assigned in early 2003. While the other crew members all help support an actual spacewalk, and there is a large team in mission control to support the crew in space, the training starts by focusing on a team of three. The two spacewalkers — EVA crew members — and the IVA crew member. As opposed to the "extra-vehicular" spacewalkers, the "intravehicular" (IVA) crew member plays a critical role equivalent to that of an orchestra conductor responsible for coordinating the many steps that are required to complete the multiple tasks of a spacewalk. Rick and I were the two spacewalkers; Tracy Caldwell was our IVA crew member. While this would be our second spaceflight, neither Rick or I had done a spacewalk, and this would be Tracy's first spaceflight. We were excited, reasonably confident and eager to learn.

NBL training starts early, with arrival around 6:45 in the morning. There is a review briefing of the tasks and crew coordination, followed by a medical examination to determine if both EVA crewmembers are "fit for duty." A head cold, the flu or seasonal allergies would preclude working in the pressurized suit. After donning the long underwear and liquid cooling garment, the two EVA crew members get into their spacesuits that are mounted on donning stands. Tracy and our commander, Scott Kelly, were there to help Rick and me get into our spacesuits as they would later do in space. Before putting my helmet on, Tracy looked me in the eye and said, "Have a good [training] run. We'll see you in a few hours." She put my helmet on, it was latched, secured and locked. I felt alone.

In the early training sessions, we skipped over the many steps involved in depressurizing the airlock. Those would be trained in a different simulator and incorporated into underwater training at a later stage. After our weigh-out to ensure we were neutrally buoyant, our divers put us in the airlock head-to-toe, the only way we fit, so we could practice getting out of the airlock without getting ourselves tangled up in each other's safety tethers. This time Rick would egress head first as the first crew member out and I would egress feet first as the second. "You're go for egress," Tracy said to Rick, her voice a calm reassurance that we had this, that things would go well. Rick moved slowly out of the airlock into position and said, "Dave, I'm tethered to the aft airlock handrail and ready for you to egress." "Thanks," I replied, thrilled to be starting our training. For the next six hours we completed all our tasks and did a pretty good job of staying on the timeline. Occasionally we would fall behind, then we'd catch up, and every now and then we would actually get ahead on the timeline. After it was over, Rick and I let the divers position us on the donning stand that was also used to lift us into and out of the water. I felt tired but pleased with how we had done.

Debrief

After getting showered and changed, Rick and I went upstairs to the control room overlooking the pool. Tracy and our training team had been on console in that room the entire time Rick and I were underwater. There were communications consoles and numerous TV monitors to help Tracy and the trainers see what we were doing while we were following the complex procedures associated with the tasks we were performing. Paul Boehm, STS-118 spacewalking lead, started the debrief, asking Rick and me, "How do you think that went?" We both looked at each other, having already chatted while changing for the debrief, and answered, "We think it went pretty well. There were a couple of times we got behind on the timeline, but we got caught up and overall it seemed like a good run." Paul smiled and looked at Tracy, "Tracy, what do you think?" Tracy paused before she spoke; with Rick and I both looking at her, she said, "Well, if we do in space what we did today, it may not work very well." Rick and I were surprised — from our perspective it seemed to have gone well. Why was Tracy concerned?

Paul, continuing to smile, said, "What do you mean?" With Tracy's answer we came together as a team. "Well," she said, "Dave, you egressed the airlock without letting me know and then once you were both outside you both went to your worksite and started the tasks. I was following what you were doing but had little insight into how things were going. You need to give me more feedback so I can help you." She was right. We listened thoughtfully to her comments that were echoed by Paul and the rest of the training team. From a technical perspective we had done pretty well. From a team perspective, we had to learn to trust each other, to communicate effectively, to leverage our individual skills to succeed as a team. For us, that was a pivotal moment in coming together as a team.

Not all spacewalkers have effectively used their IVA crew member, leading to loss of efficiency in space or, in some cases, loss of tools or mistakes that have mission impact. We did not want to be like that. In his book *Controlling Risk in a Dangerous World*,[1] five-time space shuttle commander Jim Wetherbee speaks of the importance of the "two-person" rule in reducing the possibility of error. While there is unquestionable merit in reducing error by decreasing the number of mistakes through training, operational organizations understand that "to err is human." Humans are not infallible. There are many reasons mistakes are made and decreasing the probability of error through training is an appropriate strategy. However, it is not the only strategy. Occasionally, highly trained, competent individuals make mistakes. Experienced teams are able to eliminate the consequence of individual error by leveraging the input of other team members to trap or catch errors before they have consequences. Every operator understands the need to build competency to reduce the risk of personal error. From all of his experience as a naval aviator and space shuttle commander, Wetherbee learned that "you need both the competence and the humility to know that you could make a mistake and not let yourself make that mistake. Somebody explained it to me one time. . . . That's what makes great operators — they're ones who do have the skill and confidence. But what drives them to perfection, or to strive for perfection, is what I used to call self-doubt, healthy self-doubt . . . it causes you to think about the decision much more deeply than other leaders. That's what makes you successful."[2]

Optimize

While Rick and I were confident in our skills, we were also humble enough to understand we could make mistakes. Tracy was right; we could leverage the role of the IVA crew person to ensure that we had the best chance of success in completing the tasks for each spacewalk. The next training run was different. We were a tightly coordinated team, effectively communicating what we were doing, and with Tracy's guidance through the complex procedures and oversight of the timeline, we were able to complete of our objectives on time without mistakes. There were still a lot of learning opportunities as we tried to improve the efficiency of the procedures and how we performed the tasks to ensure that we would stay on the timeline when we did the tasks in space.

Effective communication is critical and critical communication is not easy.

> "Embrace stepping outside your comfort zone, collaborate and bring in the voices of others, expand your perspective, and communicate and share what you learn along the way."
> — SCOTT KELLY

Sometimes it takes courage to speak up and say the things that need to be said to optimize success. Both the loss of Challenger and Columbia are examples of ineffective communication. Either the correct questions were not being asked, or the replies to other questions, and in some cases unsolicited input from technical experts expressing concern, were not heard. **Effective communication requires clearly** stating the issue or concern, unbiased listening to understand the message has been heard and verification that the message is understood. In a high-stakes environment if the goal is to succeed, building high-trust relationships between team members enables teams to have tough conversations to build their competency.

Tethering is critical during a spacewalk. In some ways it is like rock climbing where it is important to "make before breaking" a tether, to attach another tether first before removing a tether. We always use a redundant tether protocol at the worksite, with a local tether to hold us close to the worksite and another primary safety tether attached to either the airlock or another distant point on the space station. In some cases, like our first spacewalk to install the S5 truss element on the International Space Station, we are so far away from the airlock that we needed two airlock tethers. They are not attached end to end; rather we tether the first to a handrail that also serves as the origin of the second tether.

Safety in space is about creating redundancies, backup pathways that can be used if the primary one fails. For spacewalkers, the tether is a fail-safe that can prevent the loss of equipment or in a worst-case scenario the risk of floating freely in space if one were to come off structure. Each space suit has a Simplified Aid for EVA Rescue (SAFER), a jet pack that can be used to fly back to safety. Although astronauts are highly trained to operate the SAFER, no one wants to inadvertently come off structure. We worked closely as a crew to ensure that our tether protocol would prevent any misadventures in space. Using the "two-person" rule, we learned in the NBL how to confirm our tether actions with another crew member without slowing us down on the timeline.

Focus

For each spacewalk, we trained a minimum of a ten-to-one ratio, practicing each spacewalk at least 10 times. This was in part due to the delay in our launch following the tragic loss of the Columbia crew, but also due to the complexity of the spacewalks we would be performing. Given the potential for fatigue to contribute to human error, I intentionally stayed up late before a few of the NBL training sessions to see what it would be like to do a spacewalk when I was tired.

As is the case with many operational tasks, focus is critical. Whether you're a surgeon, a Formula One race car driver, a pilot, an astronaut or an operator in a different setting, the ability to focus is closely tied to safety and successful outcomes. Reduced ability to focus was one of the first effects of fatigue that was evident when I started the NBL training session for our second spacewalk with four hours' sleep the night before. Vigilance and focus took extra effort. Later, I was glad that I had tried training while fatigued, as Rick and I were woken the night before our first spacewalk by a space station alarm. I knew what to expect when we went out the hatch a few hours later, and I was glad that I had created that learning opportunity for myself. The spacewalk went very well, with Rick and me completing all of the tasks and a couple of get-ahead tasks.

The complex choreography of the second spacewalk to replace the failing gyroscope and correctly attach the many electrical connections normally required complete focus for success. Every detail of the spacewalk was critical. This was the perfect training session for me to evaluate the effects of fatigue on my performance. To maximize the training benefit of the NBL sessions, we train like we fly and fly like we train. While the NBL enables us to improve the efficiency of procedures, it also provides a safe environment to learn from errors to prevent mistakes in space while developing our team skills to back each other up and catch any errors before they have operational consequences.

That day I learned about fatigue, focus and great teamwork to trap errors. We were at the point in the spacewalk where I was riding the space station robotic Canadarm2 towards the payload bay to pick up the new gyroscope. Charlie Hobaugh, call sign Scorch, was the arm operator and he did an outstanding job positioning me by the new gyroscope. My safety tether was attached to the foot restraint on the end of the robotic arm, a technique that we used to ensure we would remain attached to the arm if we came out of our foot restraint. Despite the efficiency of the heel lock feature of the restraint, which we engage by turning our heels out once we're in the restraint, it is not unheard of for a heel to inadvertently come out. With a tether it would be relatively straightforward to get back in, although in my case this would not be an option while holding onto a 1,200-pound gyroscope and its mounting bracket.

Before Rick and I removed the bolts attaching the gyroscope and its bracket from the stowage platform, I attached a tether to the gyroscope and a local tether to the handrail on the sill of the space shuttle's payload bay. Everything was proceeding like clockwork despite my fatigue, but it required all my effort to stay focused. Once we removed all the bolts, it was time to move slowly away from the stowage platform, after which Scorch would move me back towards the space station. "Go for motion," I called, both hands firmly holding the gyroscope. Just as the arm began to move, I noticed my local tether was still in place. "Stop motion," I called. Needless to say, that got everyone's attention. Scorch asked, "What's up?" "It won't take long," I replied. "I have to release my local tether." The gyroscope was still stable against the stowage platform and I had a tether on it as well, so I was able to use my left hand to stabilize the gyroscope and with my right hand I quickly detached my local tether. Fatigued, I had not noticed this before I called for motion. We debriefed it after the training session, and I committed the verification step to memory to ensure it wouldn't happen again in training or in space.

Affirmative

After my first spacewalk of the mission, I wrote in my crew note-book, *Fantastic spacewalk! It went as planned despite being woken up in the middle of our sleep period. Great illustration of train like you fly and fly like you train.* Heading into our second spacewalk, our teamwork continued to keep us error free and on the timeline. I marveled at the view as I rode the Canadarm2 towards the payload bay to get the new gyroscope. This time, I double-checked each step of my tether protocol before calling for motion. When I called for motion, as a double-check, Tracy asked, "Local tethers clear?" I responded, "Affirmative," and Scorch moved me out of the payload bay back towards the space station. That was teamwork. **It's not about egos and worrying whether or not someone will get upset if you remind them of something, it's about making sure that we're all working together to enhance safety, efficiency and success.**

Now, with a face full of gyroscope, it was impossible to see the spectacular view of the Earth far beneath me. Despite the many risks associated with manually moving a 1,200-pound object with two hands while perched on the end of a robotic arm in space, Scorch called, "How's the view?" He knew I couldn't see anything except gyroscope. His wry sense of humor brought a smile to my face and made me think about teamwork, optimizing success and managing emotional energy. Great teams know when to use humor, and we succeeded because we had come together as a great team.

Building peak-performing teams takes time. It also takes com-mitment. Everyone on the team must commit to ensuring they are using their skills to actively participate in team activities. Individual competencies are as critical as team competencies and, while the first creates the foundation for the second, it is often how teams work together that determines success. In corporate teams, individual skills, collaborative skills, team size, mandate, resources, organizational culture and reporting structure are all important. Putting together a spacewalking team is a subset of assigning a crew to a mission. In a high-stakes environment, aligning individual technical competencies with tasks is an important first step enhanced by selecting individuals with behavioral attributes that will optimize team success.

Unlike technical skills where a team member may be selected for their expertise in one particular area, effective team members have a breadth of non-technical skills including proven collaborative skills, communication skills, negotiating and conflict resolution skills, leadership and followership skills as well as a willingness for hard work, patience, commitment and the ability to enhance the emotional energy of the team. Many of us have had the experience of working with technically competent individuals who "suck the life out of you." While those individuals may be selected for their superior technical skills, their potential negative impact to the team can be significant. Team members with stronger team skills can enhance technical skills through training. It is often more difficult to change the behavior of a difficult team member.

Diversity

> "I thought it was important to search out the candidates across the country and give everyone an opportunity to apply to the program." — GEORGE ABBEY

In a competency-driven environment, selecting solely on the basis of technical competency decreases the diversity of a group. It is possible to have both. Stronger teams are created when, in addition to selecting for technical competency, members from different ethnicities, cultures and genders are selected. George Abbey recognized this in hiring the first group of astronauts for the Space Shuttle program in 1978.[3] Of the 35 astronauts selected, six were women, three were male Black Americans, and one was a male Asian American. At the time this was a significant first step in building diversity within the astronaut corps, an approach Abbey continued in subsequent selections. Unsuccessful but promising candidates were given suggestions on what they might do to improve the likelihood of success if they applied again and some were offered technical jobs at Johnson Space Center to help them build their skills. Irrespective of technical background, a willingness to work collaboratively, practice team skills and an ability to learn a breadth of new technical skills ultimately predicted those who would perform best in space. There is no question that spaceflight is the ultimate team activity and arguably spacewalking is the epitome of that saying.

Great organizations embrace diversity as they create great teams through a commitment to continuous learning and talent development. Those are the organizations that we want to be part of, where a vibrant culture with a compelling mission that attracts the best grows stronger through proactive development. The thousands of individuals in the large pyramid supporting each of our spacewalks were part of that culture and as a team we used everyone's skills to achieve success.

Leadership Insights

- The "two-person" rule reduces both the possibility and consequences of error.
- Communication is critical and critical communication is difficult.
- Prepare for what you will be doing and do what you have prepared for.
- Building teams takes time, commitment, communication and trust. Peak-performing teams are typically high-trust teams.

CHAPTER 19

Become a Listener

*"We shouldn't abbreviate the truth but rather
get a new method of presentation."*
— EDWARD TUFTE

Gazing over his round glasses, Edward Tufte looked out at the audience, speaking slowly to emphasize his point. "Evidence is evidence, whether words, numbers, images, diagrams, still or moving. The information doesn't care what it is, the content doesn't care what it is, it is all information. And for readers and viewers the intellectual tasks remain constant . . . to understand and to reason about the materials at hand, and to appraise their quality relevance, and integrity."[1]
Tufte is a professor emeritus of political science, computer science and statistics at Yale University. He is passionate about the visual display of quantitative information. He has written a number of books with such varied titles as *Seeing With Fresh Eyes*, *Beautiful Evidence* and *The Visual Display of Quantitative Information*. He is a uniquely modern blend of statistician and artist. He is an expert on data analysis when truth matters. He was also one of the many experts asked to present to the Columbia Accident Investigation Board (CAIB).

The morning of February 1, 2003, the families of the astronauts on STS-107, the 28th mission of the space shuttle Columbia, were patiently awaiting the return of their loved ones at Kennedy Space Center in Florida. Unlike the fanfare of a shuttle launch, landings seemed almost anti-climactic. There were typically a handful of media to cover the post-flight press conference from the crew, perhaps a few curious onlookers on the causeway interested in seeing a shuttle landing, and the NASA team ready to assist the crew off the vehicle and begin processing the orbiter for another mission. The families stood in anticipation near Runway 33 waiting for the shuttle to land. It had been an early morning getting ready for the scheduled touchdown at 9:16 a.m. EST, but everyone was wide awake anticipating the hugs and congratulations that would soon come. The families of the veteran astronauts listened carefully for the twin sonic booms characteristic of the space shuttle transitioning to subsonic flight as it approached the space center. That morning, the sound never occurred.

At 8:59 a.m., just 17 minutes before the expected landing, mission STS-107 ended with the break-up of the spacecraft over northeastern Texas. By all accounts it had been another successful scientific mission with a number of experiments and commercial payloads onboard. Yet the tragic ending of the mission was the result of an event that took place 81.7 seconds after liftoff on January 16, 16 days earlier. During launch, a piece of foam separated from the left bipod ramp of the external tank of the orbiter, hitting the leading edge of the shuttle's left wing at roughly 545 mph. The foam strike was not evident during the launch; from the mission control and crew perspective it had been another incredible eight-and-a-half-minute ride to space. However, within 24 hours the post-launch review of the launch video imagery discovered evidence of the impact. Frame-by-frame analysis of the video revealed a shower of material from the wing. The critical question on everyone's mind was whether the impact had damaged the orbiter and if so, what was the extent of the damage. It would be up to the Mission Management Team (MMT) to answer the question.

Mission support is a 24/7 activity. There are teams of experts on console throughout the mission to monitor spacecraft systems and communicate with the crew. Liftoff and landing, the dynamic phases of flight, are associated with the highest levels of risk but the vigilance continues throughout the mission as experience has shown that anything can happen at any time. During a mission there are daily MMT meetings where senior program managers meet to discuss any issues relevant to mission success. Engineering data is reported to the MMT by the manager of the mission evaluation room, a support function of the Space Shuttle Program Office to give engineering and technical feedback during a mission. The chair of the MMT provides mission status updates to the shuttle program manager, who in turn ensures that NASA senior management is informed of relevant issues.

The video imagery resulted in immediate action by a debris assessment team with input from Boeing and NASA engineers. While foam strikes had occurred before in the program, the estimated size of the foam that hit Columbia was greater than anything that had been seen before. The crew had sent some video of the external tank after it separated from the shuttle, but the area of foam loss was not visible in the images. While the assessment team worked with the imagery they had, discussions began about using military assets to provide images of the Columbia's left wing during the mission. Concurrently, the Boeing team in Houston used a software program called Crater to model the results of the impact. Crater was designed for "in-family" impact events and was intended for day-of-launch analysis of debris impacts. It was not intended for large projectiles like those observed on STS-107.[2] Based on the results of the Crater modeling, the previous launch experience with other episodes of foam loss and flight readiness review decisions about the risk of foam impacts to the orbiter, the MMT concluded that the foam impact to Columbia's wing would not endanger the orbiter or crew during reentry. That conclusion and the ensuing events were a tragic reminder of the risks and challenges of flying one of the most complex spacecraft in history.

NASA is an organization of extremely talented, technically competent, dedicated personnel. Wayne Hale had been a flight director at Johnson Space Center for 15 years before his promotion to becoming the manager of shuttle launch integration at Kennedy Space Center. He was to start work February 1, 2003, the day Columbia was scheduled to land. He had packed up his car and driven to Florida, arriving the day before landing, and was at the landing site with other NASA personnel to welcome the crew home. That eager anticipation quickly turned to profound grief and a gut-wrenching sense of failure that spread throughout the agency. "I mean, I thought our organization was great. I thought we could handle anything," he said during an NPR interview in 2017.[3] But like others, he was aware of the problem of repeated foam strikes on the orbiter and based on the data, discussions and shared experience of the team, "We all felt pretty good. This was not going to be a safety issue."[4]

Hale never got to work as the manager of launch integration. After the loss of Columbia, Bill Parsons became the new shuttle program manager and later that summer called him and said, "[I'd] really like you to come be the deputy program manager in Houston." Hale would spend the next two and a half years working with the team getting the orbiter back to flight status. **It is what people do when failure occurs that determines whether they will ultimately succeed.** Hale realized that there needed to be a change in the NASA culture, that experience and available data had prevented the team from "seeing with fresh eyes." NASA leaders needed to be better at the art of communication. Team members and leaders had to improve how they voiced and heard their concerns. Hale recognized that effective communication was as important in controlling risk as the most complex engineering analysis and perhaps the biggest change for everyone at NASA was to become better at listening.

Retired U.S. Navy Admiral Harold W. Gehman led the seven-month investigation into the loss of Columbia. Within two hours of loss of signal from the spacecraft, the CAIB was being formed. Over the next few months, its 13 members were supported by a team of more than 120 individuals along with 400 NASA engineers. They reviewed more than 30,000 documents, conducted more than 200 interviews, heard testimony from numerous expert witnesses and reviewed more than 3,000 inputs from the public.[5] As in the case of Challenger and perhaps Apollo 1, the board recognized that the organizational culture within the human spaceflight program played an important role in the tragic loss of the vehicle and crew. "The board's conviction regarding the importance of these factors strengthened as the investigation progressed, with the result that this report, in its findings, conclusions and recommendations, places as much weight on these causal factors as on the more easily understood and corrected physical cause of the accident."[6] In many ways those factors were similar to those described by sociologist Diane Vaughan in her book *The Challenger Launch Decision: Risky Technology, Culture and Deviance at NASA.*[7] Once again she described what happened as the social normalization of deviance: "[It] means that people within the organization become so accustomed to a deviant behavior that they don't consider it as deviant, despite the fact that they far exceed their own rules for elementary safety."

No one comes into work thinking that it is acceptable to break rules. Powerful external forces are at play in shaping those decisions. Overt schedule pressures affected the Challenger launch decision and undoubtedly the Columbia team were similarly concerned about how the issue of foam strikes would affect future launches to continue space station construction and operations. Perhaps one lesson that applies to every organization is the subtlety of decision creep in the evolution of deviance and how even the best organizations are not immune. It challenges all senior leaders to consider how decisions are being made within their organization and the subtle forces of social pressure, choice, conformance and expediency. Individuals who speak up often do so at their own risk, yet rather than being chastised, their courage in challenging the status quo should be welcomed. There are those willing to do the right thing even if it costs them their job, but for most it is easier to simply "go with the flow." In the arena of high-stakes operational activities, effective communication plays an important role controlling risk.

Not surprisingly, there are different ways of communicating within organizations. Critical information or data should be shared within teams and with leaders using critical communication skills. In commercial aviation, pilots are required to read back instructions from air traffic control to verify they have been heard and understood. Similarly, critical communication involves speaking up to concisely share information, the implications of the information and any recommendations in a manner that is easily understood. The listener has the responsibility to listen up, understand the message and if necessary, paraphrase what they have heard back to the speaker. It is a skill that takes courage. It is difficult to share complex information that tells the listener what they need to hear, not what they want to hear. Communicating concerns and ensuring they are heard was an issue in the 1986 loss of Challenger before the widespread use of email and other computer applications characteristic of the modern workplace. The reliance on engineering by viewgraphs caught the attention of the CAIB and they turned to Tufte for his insights.

Tufte focused on Slide 6 in a PowerPoint presentation prepared by the Debris Assessment Team that was used to communicate the results of the Crater modeling test data. His analysis reveals the use of ambiguous phrases with vague, confusing, "sloppy" language that may have contributed to the approval by the decision-makers that the vehicle was safe for reentry. The approach of NASA leaders like George Low, who was renowned for his probing questions during the verbal briefings from engineers, had transitioned to a culture that used electronic briefing slides instead of technical papers, discourse and debate for technical communication. One of the bullet points on the slide stated that "significant energy is required for the softer SOFI [spray on foam insulation] particle to penetrate the relatively hard tile coating." This point was supported by a sub-bullet that in retrospect is very concerning: "Test results do show that it is possible at sufficient mass and velocity."[8] How did a room of highly competent technical experts not focus on the words "it is possible" and ask what mass and velocity would cause damage to the vehicle? Perhaps reassuring words such as "conservatism" in the slide title decreased their vigilance. In the realm of complex modern organizations where time is always at a premium it is understandable that what a manager might interpret could have been influenced by the title alone. The board found that "risk information and data from hazard analyses are not communicated effectively to the risk assessment and mission assurance process."[9] They also were surprised and concerned when they received similar presentation slides instead of technical reports from NASA officials during their investigation. There is clearly a role for electronic presentations, but it is critical that they effectively communicate the data that will be used to make decisions and are supported by in-depth technical analyses.

The board did not comment on the use of email, a form of communication that is now ubiquitous in organizations, as a tool to share critical concerns. Like many tools, email can be extremely effective if used correctly. Yet it is easy to hide behind an email chain instead of having a courageous, face-to-face conversation. There are approximately 250 working days in a year and the average senior leader typically receives 50 to 75, or more, daily emails. Instead of someone directly speaking with them about an issue, leaders often have to struggle to find critical bits of information in the 12,000 to 18,000 emails that pass through their inbox every year. In some modern-day offices, people quickly learn to "broadly CC for accountability and BCC to CYA," generating packed inboxes and the need to sort through a lot of information to find important points. A number of emails to Columbia managers went unanswered and, whether they were seen, read or buried in an overflowing inbox, it suggests the need for a different approach to communicating critical information. **Despite the benefits of modern technology, the best approach is person-to-person verbal or face-to-face discussion.** Mike Griffin, the NASA administrator tasked with returning the shuttle to flight after the loss of Columbia, chose to make himself available to discuss any area of concern.

Griffin's wife Becky recalls getting a call from Marshall Space Flight Center Director Dave King very early in Griffin's tenure. King had called the house on the weekend looking for Griffin. She recalled, "Dave will tell you that he was probably more nervous calling our house than just about anything else. I said, 'Don't ever worry about it. If you need Mike, you need Mike. I don't care what time of day or night you call, call if you need him.'"[10] The actions of leaders are critical in improving or shutting down communication. **What a leader says when they get the call on a weekend, evening or the middle of the night will determine whether or not they continue to be informed of important events in a timely manner.** Leaders like Griffin welcome the calls.

When Griffin took over, the agency was preparing the shuttle for the STS-114 return to flight mission. Griffin wanted NASA to return to flying the shuttle, even though he had been concerned with its design in the past. He wanted to safely return to flight to finish building the International Space Station and keep the commitments NASA had made to its international partners. He had the support of the White House. President Bush strongly agreed that the immediate priority was to finish what had been started and then look beyond Earth orbit back to the Moon. Griffin wanted NASA to go back to the Moon with a mission architecture that would work. The first step on that journey would be safely resuming shuttle flights.

As the administrator ultimately responsible for human space-flight, and as an engineer, Griffin wanted to be kept informed of key milestones in the return to flight timeline. Within a week of his arrival at the agency he gave direction that he was to be invited to every debris review meeting. Shortly after, there was a critical meeting in Houston that he found out about at the last minute. He flew down and walked into the meeting with roughly five minutes to spare. That was the last time he was not informed of a meeting.

During the meeting, he was concerned that people were willing to sign off on what he regarded as an incomplete technical assessment of ice formation and shedding from an area of the shuttle's external tank called the lox feedline bellows. As an experienced leader in the aerospace sector, Griffin knew that **it is critical to fly when you're ready to fly, not necessarily when you're scheduled to fly.** His concern grew when he realized that "they were, by their own admission, a couple of months away from having a complete analysis and yet, the senior managers were willing to fly."[11] Looking slowly around the room, he told them they were not going to fly until they were ready. He told them that he preferred they take another month and a half or two months to get a complete analysis done. Given that extra time, Griffin insisted on implementing a recommendation made by Shuttle Program Manager Bill Parsons to put a heater on the liquid oxygen feedline bellows, the area on the external tank where foam loss had occurred in the past.

STS-114 launched July 26, 2005, at 10:39 a.m. EDT, 29 months after the loss of Columbia. The mission delivered supplies to the International Space Station and the crew successfully used the new orbiter boom sensor system (OBSS) that was installed on the end of the Canadarm to inspect tiles on the undersurface of the orbiter. The mission objectives were accomplished, but unfortunately a large piece of foam came off the tank 127 seconds into the flight. The foam did not hit the vehicle, but it was unexpected after all the work that had gone into assuring a safe return to flight. Post-flight review of the anomaly indicated the foam loss resulted from a repair that was made at the Michoud Assembly Facility. Information on the repair and re-gluing had not been made available to senior management. Griffin chose to ground the fleet, a decision about which he said, "That created a lot of flak in the press about how we were never going to finish the station. That one went all the way to the president, who backed me. The words that came back to me were 'whatever Mike says we need to do, that's what we'll do.'"[12] Griffin knew it was the right decision even though within three months of being on the job, he felt his job was on the carpet because at the White House and congressional level, grounding the fleet was not welcomed. But the president continued to support him, and the fleet stayed grounded until July 2006.

Griffin did not shy away from the decision to do what was right. "As the leader, I know that I'm being judged by the decisions that go out the door. Not who made the decision, not whose idea it was. And that's something I've internalized decades ago. I'm judged, I'm rated, my performance is graded by those who grade me, not by whether or not I thought up the idea, but whether the idea that went out the door is the best one. So, I try to tell people that it doesn't matter where the idea comes from, or whose argument eventually carries the day. **From my perspective as the boss, all I want is the best idea and it doesn't have to be mine.**"[13] During the interview for this book, his passion for role modeling the leadership and team skills he thought were critical to successfully returning to flight was immediately evident. He emphasized, "I never hesitate to apologize when I've been wrong, or to admit when I've been wrong. I'll say, okay, I was wrong on that. Let's move on. So, if the boss says, 'My mistake, you know, we were focusing on the wrong thing, let's move on,' then everybody else can breathe a sigh of relief because they're not being blamed. **The goal is to get a good decision out the door. You can't do that if everybody's obsessing over who made a mistake."**

Probably the most significant mentor that Griffin had was his former boss, Lieutenant General Jim Abrahamson, who was the associate administrator for the Space Shuttle program at NASA headquarters during the early shuttle flights. For the five or six years that he worked for Abrahamson, he was able to watch how he led and ran things. Griffin described him as "the best leader of people I ever observed. I used to say, I still say, that everything I know about leadership, I learned from Abe. My only regret is I didn't learn everything he had to teach." The skills Griffin learned from Abrahamson were critical in ensuring a safe return of the space shuttle to flight for the remainder of its career.

Tuesday, February 4, 2003, was a tragedy for those involved in human spaceflight and the families and loved ones of the seven crew aboard Columbia. Commander Rick Husband, Pilot William McCool, Payload Commander Michael Anderson, Mission Specialist David Brown, Mission Specialist Kalpana Chawla, Mission Specialist Laurel Clark and Payload Specialist Ilan Ramon were mourned in a special memorial ceremony at Johnson Space Center. As President George W. Bush addressed the group, he shared a conversation that had taken place between Dave Brown and his brother a few weeks before launch. His brother had asked what would happen if something went wrong on their mission. Brown replied, "This program will go on." The president emphasized, "Captain Brown was correct. America's space program will go on. This cause of exploration and discovery is not an option we choose; it is a desire written in the human heart. . . . We find the best among us, send them forth into unmapped darkness and pray they will return. They go in peace for all mankind and all mankind is in their debt."[14]

There are no words to describe the grief that was felt by the families and friends of the astronauts. The entire NASA community shared their pain. The day before the memorial service, I (Dave) had returned to Houston from Lufkin, Texas, where I had been participating in the recovery effort that had been ongoing since the loss of the crew and vehicle. Three days before, on that fateful Saturday morning, our family had been watching the landing on television with some friends who were visiting from out of town. As the tragedy unfolded, the phone rang. It was Dr. Rich Williams, NASA's chief medical officer at NASA headquarters in Washington. "Are you watching the landing?" he asked. "Yes," I replied, "I'm getting ready to go to JSC and I'll call you when I get there." It seemed as though my car was on autopilot during the familiar 10-minute drive.

I started by going to my former office on the eighth floor of the administrative building and had a brief chat with Dr. Jeff Davis, my successor as director of the Space and Life Sciences Directorate (SLSD). He was on the phone discussing the plans that dealt with the loss of a crew and vehicle that had been revised for me before I had been reassigned to train for a new spaceflight a couple of months earlier. He briefly put the phone on mute, and I asked, "Is the plan working, do you have any questions for me?" He responded, "No, we're good. If I have any questions, I'll give you a call." "Sounds good," I replied, "I'm heading over to the astronaut office to see if there is anything I can help with." Two hours later I was in a NASA aviation safety truck headed for Lufkin, Texas, with Jim Wetherbee, the former director of the Flight Crew Operations Directorate (FCOD) at JSC and two fellow astronauts, George Zamka and Barry Wilmore.

Jim had been through this before following the loss of Challenger 17 years earlier. At that time, George Abbey had been in charge of flight crew operations and played a key role in the recovery efforts. "Has anyone spoken with George yet?" I asked Jim. "I don't think so," he replied. I dialed the number for him and passed him my cell phone.

Readiness is critical when working in high-stakes operational environments. I recalled a meeting with Dan Goldin, Jim Wetherbee, Charlie Precourt (then head of the astronaut office), Senator Glenn and his fellow astronauts in the crew of STS-95. Goldin had asked, "What is this operational environment that you keep referring to?" Precourt responded, "It is a time-critical environment where decisions have life-or-death consequences and if you make the wrong decision you can't take it back. You can only modify it with subsequent decisions." Astronaut training, simulations with mission control, continuous learning and practice are all about being ready to deal with whatever might happen. The loss of a spacecraft is something that is difficult to prepare for. It is preparing for a catastrophe, a tragedy beyond description, the loss of friends and colleagues, knowing that their families will never be the same. It is preparing for something you hope will never happen. In the last six months as director of the Space and Life Sciences Directorate, I had asked Dr. Phil Stepaniak to revise the plan. He had done an outstanding job finishing the comprehensive plan in the fall of 2002. We hoped we would never have to use it, but we were ready. Neither of us imagined that within three months we would be putting the plan to the test.

After five minutes, Jim finished his call with George Abbey. We didn't have time to discuss it as my phone rang again, then another call and another. I would receive another 50 calls before the day was over. Before each spaceflight there is a certification of flight readiness by each part of the program supporting the mission. As director of the Space and Life Sciences Directorate (SLSD), I had been part of the Mission Management Team and sat with other members of senior management on console in the launch control center at Kennedy Space Center. Every member of the team became very good at having with them everything they would need if there were any problems. Today we would have all of our documentation available electronically on a smartphone, tablet or computer but in

2003 we were still using a combination of electronic and printed information. In the SLSD, we had prepared a small pocket-sized booklet of all of the information that would be needed in the event of a contingency or loss of the vehicle. After supporting over 20 shuttle missions, I was ready and had my go-kit with me in the van: cell phone, laptop, chargers, clothes and my contingency reference booklet. Over the next three days my copy was in continuous use by our mortuary team. At the command center copies were made for each of the leaders. It was the primary source of contact information, crew data, hazardous materials, orbiter payload and other information for the first 24 hours.

I will never forget the time I spent in northeastern Texas. I was in and out of the field between February and June 2003, and while the loss of Columbia was the most tragic event of my career, the response of the local community, the U.S Forestry Department volunteer searchers, FEMA, the NASA teams and the many branches of law enforcement, the EMS and firefighters was the finest teamwork I had ever seen. Return to flight would be based on understanding what had happened. Recovering the vehicle and crew were critical, urgent elements in the path back to space. To me, the recovery was America at its finest.

It was well after midnight, in the early hours of February 2 that it was finally time for bed. One of the FBI evidence response technicians asked me where I was staying. I thought about it for a minute and said that I would be staying where I was. For me, leaving was not really an option. I was able to put together a makeshift sleeping area on the floor of the command center with a borrowed blanket and pillow. It seemed appropriate that someone from the astronaut office be with the crew. The sadness I felt was like a weight that was crushing me. It felt a hundred times worse than the 3 g's I had experienced during my last launch to space. The quiet and darkness surrounded me and I resolved that I would continue the mission, to do everything I could to help with the recovery, to train as hard as I could for my next mission STS-118 and to get back to space to continue the quest of exploration shared by my fallen friends.

Leadership Insights

- Listening is more important for leaders than speaking. Understanding rather than simply hearing the message is essential.
- When confronted with divergent opinions and disagreement it is safer to identify the most conservative thing to do, then do it.
- Organizations, like the people in them, respond with grief at times of significant loss. Leaders should consider both the emotional and technical aspects of recovering from failure.

CHAPTER 20

That Didn't Work. What's Next?

"No matter what you do, it's like whack-a-mole."
— JOE ROTHENBERG

"You walk in in the morning and you've got three problems that are going to bring [down] the whole project — whether it be the Hubble repair mission, the Space Station, or building these solar arrays — how am I ever going to solve that?"[1] Managing the technical challenges of large complex programs can be daunting, recalled Joe Rothenberg, as he described the challenge on his hands with the Hubble Space Telescope in the early 1990s. "By lunchtime, two of the unsolvable ones are solved. The other one looks like it might be solved." Persistence and optimism are necessary when working in the unforgiving vacuum of space.

The associate director of flight projects for the telescope was trying to fix the problem bedevilling the agency: the mighty telescope had launched with humbling astigmatism. Two months after launch, Hubble chief scientist Ed Weiler held a press conference to announce that the $1.5 billion telescope's mirror was unable to provide a sharp image. The most celebrated telescope in history was now center stage in the arena of media ridicule, and the out-of-focus images created outrage with Congress. The team had to find a solution.

Hubble was already late and over budget when it launched on Space Shuttle Discovery on April 24, 1990. Part of the overrun came down to bad luck; the loss of Challenger four years before delayed every flight by two years, including Hubble's. The soaring budget was tied to the usual challenges with new engineering projects working with novel technologies that must perform in that most demanding of environments, outer space. The telescope mirror in particular was a challenge. Its engineering was so precise that the companies building Hubble used computer-controlled polishing machines to make the mirror exactly the right shape to catch the light of distant stars and galaxies.

Better late than never, NASA officials had reassured the public, and up the telescope went. Every spacecraft must undergo a commissioning period in space to iron out communications wrinkles and make sure that the instruments and other components are in tiptop shape. During Hubble's commissioning period, however, a fatal flaw emerged: the powerful polishing machine had been directed to polish the mirror to the wrong shape. Hubble was flying but it was slightly myopic, an optical distortion referred to as Spherical Abberration, and the promise of it peering back in time to learn about the universe's evolution was at risk.

Late-night comedians and newspaper columnists alike howled at Hubble's inglorious predicament. Rothenberg, however, began to attack the problem like a systems engineer. Yes, the images were out of focus and a solution had to be found. But there were a couple of things working in NASA's favor. The most ironic one was Hubble's location. The telescope was inconveniently located in space. Its orbit, however, was close enough to home to allow astronauts to reach it, to repair it, to make it right again.

Rescue

Repairing Hubble turned into a lesson in consensus building and Rothenberg was up for the challenge. He first went to the people who wanted to use the telescope — the scientists competing for the chance to use the capabilities a repaired Hubble could provide. A rescue, Rothenberg promised them, would not only fix the problem with the mirror; it could also include upgrading the instruments to make the telescope more capable.

"That helped the science community buy into the plan that we're not just going to take all 15 years of the program in front of us, take all the money and spend it for this repair,"[2] Rothenberg said. But it wouldn't be an easy road. With NASA being criticized so roundly in the press, he would have to get the political funding in play to approve the mission, and a commitment from NASA that astronauts could be safely sent outside and conduct the repair.

Consensus. That was the only way to go.

Rothenberg was used to working in teams for the benefit of a mission, even a personal one. Early in his career, he faced the common problem of getting the skills he needed to do what he wanted to do. He couldn't afford to go back to school full-time and he was enjoying working on an aircraft carrier serving in the United States Navy. So, he and his wife worked out an arrangement. For 13 years, Rothenberg went to school at night. The result was worth it. He earned both a bachelor's and a master's degree and was hired by Grumman, the same company that built the Apollo lunar lander.

One of Rothenberg's first jobs was working on the environmental control loop for the LM. He created instrumentation to look for pressure drops during the system tests. In a NASA interview, Rothenberg recalled the time that Grumman planned to connect his system to a test stand. They invited NASA and got ready for a big show. The show fizzled, however, when they discovered moments beforehand that the test stand and instrumentation system were built to different sizes. No mating would be possible that day.

Lessons

Rothenberg didn't shy away from the challenge, adding the experience to the chain of "lessons learned" every space engineer learns to embrace. "That little lesson taught me . . . a skill that turned out to be very valuable," he later shared in a NASA interview, "and that was to try to make sure I understood what I was getting into and what the other side of the interface was."

Rothenberg was working at Grumman and, as many engineers did at that time, migrated to working on NASA programs at the Goddard Space Flight Center where he helped to launch four spacecraft. This was the early 1970s, when Apollo work was winding down and other projects began to wind up. One of these newer ideas was creating a telescope that could fly above the obscuring atmosphere of Earth to see farther into space than any other telescope of its size — that became known as the Hubble Space Telescope. He became a NASA employee at Goddard in 1983 to work on the Hubble and Solar Max satellites, which were among the first on-orbit servicing missions performed by the Space Shuttle.

Back in the 1970s, NASA didn't have much spacewalking experience. Moonwalking, sure — the agency had done days of it while its 12 astronauts walked about the lunar surface. But spacewalking was much less familiar. From the Gemini missions of the 1960s, NASA knew astronauts needed handholds and tethers so they wouldn't overheat themselves trying to move around outside their spacecraft. But most of these spacewalks were done for small science experiments — not to fix up a spacecraft. Fixing Hubble would require figuring out how astronauts, working in bulky spacesuits with gloves similar to those athletes use in hockey, would be able to fix delicate instruments.

Fortunately, Rothenberg not only had a broad understanding of all aspects of the telescope but more importantly a long history of working with Frank Cepollina who headed the Hubble Engineering Team. He had already sat through the early debates about what things you could swap in, what things you could swap out. He had talked to astronauts about what they felt comfortable doing, what they felt comfortable learning. It was with almost 20 years' Hubble experience and his experience as a program manager that he went to everyone after the mirror was known to be deformed, with a simple proposal: Listen. Learn. Implement.

The first task was going back to the space engineers to figure out options. "The commissioning work determined Hubble had Spherical Aberration," Rothenberg recalled, "but we needed to find out the why and how it happened to ensure we understood what the options were for correcting it. Number two, we needed the stakeholders to both help us verify and convince the broader community that we truly understand the source of the problem. We needed to develop a plan to correct the problems and, more important — given the wide distrust of NASA — convince the stakeholders which included the scientific community, NASA senior leaders and Congressional oversight committees that we got it right and to support the plan."

"[What] we had to do is build a program to repair it, but at the same time, maintain the long-term program and that is continue to build the second-generation instruments . . ." There were hundreds of issues that had to be addressed, Rothenberg said. "I created a top ten. I said I always want to know my top ten problems and make sure everybody knows them so we all know where we should focus our energies." He had a plan; the next step was getting support.

Adaptation

He started by convincing the media that Hubble was worth the fight. Rothenberg acknowledged that the mirror was flawed but pointed out it was still providing useful images. "It became clear to me right away that everybody's telling us what this can't do. But what can it do? I mean, we . . . have it up there until we fix it, what can it do? And that's important."

> "If you don't know something, you just tell somebody and then you figure out where you are." — BILL GERSTENMAIER

Working with the scientific community, he gained their trust while trying to find a temporary solution. They worked to determine "what was wrong with it, then we put in a program to measure and verify what was wrong with it that was consistent. . . . We found out you could have some ground software to make a fairly good correction." The images that resulted were "spectacular" despite the flaws in the system. He slowly began to shift the media's attention to highlight what was working. "They started to change from the first whole paragraph being about the flaw to, 'In spite of its flaw—.' It started to change, little by little."[3]

Then there were the politicians. NASA is funded by the U.S. government, which means the agency must fit into government priorities to get the money it needs. The early 1990s were tough on economies around the world. Everyone was in recession and, moreover, the U.S. was starting to get involved in the Gulf War. Recognizing that consensus takes time, Rothenberg made it a goal to talk to people every week.

"All I did was lay out the strategy, lay out the promise and invite them to have representatives [at meetings]," he said. His "Top 10" list was circulated every week to the people he wanted to talk to — those on the science working group, those in Congress, those in senior NASA management. Everyone had an opinion. Rothenberg chased down every accusation, from water vapor in the instruments to those who tried to dismiss him offhand with the usual "waste of taxpayers' money."

"I couldn't let that song go," Rothenberg said. He buttressed his arguments with even more consensus. He brought in external review teams. He created trust with highly regarded people in NASA willing to go to bat for him, including Apollo 10's Tom Stafford and NASA Administrator Dan Goldin. As senior NASA leaders, Stafford and Goldin brought in even higher standards than Rothenberg proposed, and he met them.

At Goldin's recommendation, Rothenberg had teams create risk profiles to better understand the impact of the proposed fix — the Corrective Optics Space Telescope Axial Replacement or COSTAR. Like eyeglasses on a person with astigmatism, COSTAR would subtly correct the flawed optics on Hubble. Precision would be needed, and trust would be required on everyone's part to make the repair mission fly.

Repair

Ultimately, it would take a highly trained astronaut crew and mission control team to successfully repair the highly sensitive instrument. This wouldn't be spacewalks that tested jetpacks or launched satellites from the shuttle's payload bay. Instead, it would be a delicate operation of retrieval, replacement and refurbishment. Rothenberg journeyed from Washington, D.C., to Houston, Texas, to meet personally with the head of the astronaut program, asking for the astronauts to be named at least one year in advance.

Rothenberg got his wish, and it was the dream team for Hubble repair. Every member of the crew had flown in space at least once. Every spacewalker knew EVAs and had flown in space at least twice. The team included one of NASA's best-regarded astronauts of the day, Story Musgrave, who would make his fifth spaceflight to repair Hubble. He had trained for Apollo and trained for Skylab, and when he finally got the chance to fly on the shuttle, he had tested the very spacesuit he would use for Hubble.

Although the telescope had been designed to be repaired in space, the mission would still be a unique challenge to the astronauts and mission management team. NASA needed an operationally experienced highly skilled leader to take on the role of Mission Director. Top-gun graduate and former Marine fighter pilot Randy Brinkley was up for the challenge. He had joined NASA in 1992 as a special assistant in the Office of Space Flight, before taking on the Hubble repair mission. "I just kept doing the things that had made me successful — and that was to do the best I could and surround myself with dedicated and talented people and listen to them and create an environment that would enhance their ability to succeed."[4] It was a formula that had worked for him in the past that would once again prove its merit with Hubble.

It was a long, hard road to get this far. Rothenberg found himself trying to get buy-in from everyone at NASA all the way up to the administrator, and everybody in politics up to the Government Accountability Office that keeps a close eye on how NASA spends its money.

It seemed everyone at the agency needed a voice. The weekly Top 10 lists continued, as did the endless meetings. "No matter what you do, it's whack-a-mole," Rothenberg said. "All these [moles] come out, you're working with a big rubber hammer, and you hit one down only to see another come up . . . that's project management."[5]

Finally came time to launch. On December 2, 1993, Endeavour soared into space, successfully rendezvousing and docking with Hubble three days later. The astronauts had enough time for seven EVAs if they needed it, which would almost double the four needed for the tricky installation of a new upper stage on the Intelsat VI satellite during a servicing mission in May 1992.

The spacewalkers would work in rotating teams of two — first Story Musgrave and Jeff Hoffman, then Thomas Akers and Kathryn

Thornton. Back and forth they would go, performing the tasks and communicating closely with mission control during each step of the complex repair. Again, making this telescope work would take teamwork.

Like any lengthy home renovation, Hubble's repairs came with surprises — this despite the best efforts of NASA and all its contractor partners to simulate what would happen. But the astronauts, improvising based on experience and the helpful assistance of their colleagues back in mission control, adapted and innovated. The latches on the gyro door would not reset easily, likely due to temperature changes in space. With approval, Musgrave used his weight to gently lock the latches into place. Thornton's spacesuit communications malfunctioned during her first spacewalk, forcing her to relay commands through her teammate. On a later EVA, Musgrave briefly had leak problems on his spacesuit, but managed to fix them.

The astronauts persisted and surprisingly, the most worried-about part of the complex procedures went off very well. COSTAR slipped into its allocated spot during the fourth spacewalk. By spacewalk five, everything was ready to go. Hubble was released into space, awaiting testing by ground controllers to see how far three years of work had brought the telescope.

Upgrade

Hubble may have looked the same on the outside, but the new components made it a completely different telescope. COSTAR was now in place, replacing the faulty High-Speed Photometer. The Wide Field and Planetary Camera was upgraded with the Wide Field and Planetary Camera 2. Solar arrays, gyroscopes, electrical control units and magnetometers were added or replaced. Even the computers got upgraded. And when Endeavour released Hubble, the shuttle helped the telescope boost its orbit. This would allow the telescope to operate longer in space, without the risk of falling into Earth's atmosphere.

With Endeavour landing on December 13, it was now the ground teams' turn to make sure everything was working. Over the holidays and into the new year, NASA and its partners tested the telescope thoroughly. Only one month later, on January 13, 1994, came the happy news — Hubble was ready to go. It was ready to deliver data from space.

Some in the space industry began to compare Hubble to Apollo 13, in that both missions were "successful failures." Rothenberg argues, however, that the two missions cannot really be compared beyond that. The people working on Apollo 13 had to work in constant emergency mode, getting the astronauts back to Earth within four days. Hubble, however, benefited from three years of sustained effort.

"We demonstrated that we had a competence problem, or at least a process problem," Rothenberg said about the initial Hubble failure, but gradually, he added, "we had time to fully understand the problem. We had time to double-check. We fully understood it before we made any decisions about what to do to fix it."

And three years versus three days gave Rothenberg a lot of time to build consensus. He had experts weighing in. He had independent reviews. Apollo 13, by contrast, was a "short fuse," he said because "you had little information, you had to make decisions in near real time, and human lives are at stake and you really didn't have a lot of opportunity to double down."

Rothenberg doubled down in the early 1990s and, in the years to come, Hubble would show his confidence. Just a smattering of its discoveries — and the fact that four astronaut crews were dedicated to continuously updating the telescope — will give you a sense of how important Hubble has become to our understanding of the universe.

One of Hubble's key discoveries was finding that the universe is not only expanding but accelerating in its expansion. This Nobel Prize–winning realization made everyone think carefully about how objects in the universe evolved, as well as their age. Ultimately, it may help us better understand the fate of the universe — but only after astronomers get their minds around complex features such as dark matter and fundamental particles.

Later in its career, Hubble may have saved a spacecraft. The telescope spotted several moons orbiting dwarf planet Pluto while the New Horizons spacecraft was on its way to an epic flyby through the system. Knowing where moons were positioned was the key to threading the gravitational needle as the spacecraft performed a tricky flyby. Too close to a world, or too far away, and the spacecraft could veer in a strange direction — perhaps overshooting its ability to correct through using fuel or, even more catastrophically, causing a crash with some distant celestial body. But Hubble's discovery helped New Horizons navigate safely. It made it past Pluto in 2015, whizzed past another object in 2018 called MU69 and is still sending back information that will revolutionize our understanding of icy objects far out in the solar system.

For a generation, Hubble images have been a source of curiosity and inspiration flowing across the world into textbooks, on websites and even on products such as posters and T-shirts. Millions of schoolchildren in Canada, the United States and Western Europe were shown Hubble imagery in school, and some of that "Hubble Generation" now work at NASA — some of them helping to spread Hubble's information now through social media and webcasts.

Hubble was teamwork at its finest. Commander Richard Covey said, "I could not think of any spaceflight that . . . would be more significant, more important and more fun than this one. It was truly a pleasure to . . . work with this team." Musgrave added, "I saw us doing what NASA should do: unbelievably good teamwork."[6] It was through the dedication of different teams of people that Rothenberg and Brinkley turned this telescope from being a "techno-turkey" into one of the most famous in the entire world. The "lessons learned" from Hubble stand as an example to all leaders and will surely inform the next-generation space telescope NASA is launching: James Webb. Webb will fly farther out in space and peer deeper into more distant worlds. From galaxies to exoplanets to solar system objects, our exploration of the cosmos is really only beginning.

Leadership Insights

- The time that it takes to build consensus is offset by the sustained support that comes from consensus.
- Engage a broad constituency to understand what's wrong, collaboratively develop a plan to move forward and monitor progress against the plan.
- Focus on what is working, while solving what is not working.

CHAPTER 21

Try, Try Again

"Focus more on people to prevent mistakes
and their consequences."
— BILL GERSTENMAIER

One deformed sensor transformed a rocket ride to the International Space Station into a scary abort for two astronauts.

The Expedition 57 crew performed flawlessly on that day in October 2018. They radioed the problem back to the ground. They implemented emergency procedures. The mission, cut from six months to only a few minutes, saw Aleksey Ovchinin and Nick Hague make a safe touchdown.

Within hours, they were eating lunch in front of media cameras. They were physically fine. It was a serious situation, to be sure, but the guys walked away alive and healthy.

"I think sometimes we remember the things that went really bad, but then we don't remember some of the things that really went well."[1] That's a comment that Bill Gerstenmaier, then NASA's associate administrator for human exploration and operations, once made about mission management in general. It was a comment that might have applied very well to this abort, but he and his Russian counterparts understood the importance of finding and fixing the problem. The spacecraft and rocket didn't make it far into space, the crew had been protected all the way back to Earth, yet an investigation was needed before they could try again.

Fortunately, after years of working together, there was such trust and expertise between the Russian and American teams that when the independent assessments of what went wrong were done, not only did they trust each other's findings, they discovered they had come to the same conclusions.

This launch was aborted. The next one didn't launch on time. In fact, it launched three weeks early. And no one was rushing. The schedule change was meant to smooth over crew handover procedures in orbit, giving everyone more time for discussions before the three people waiting on board the ISS came back home.

Canadian astronaut David Saint-Jacques watched the abort while in Kazakhstan training for his upcoming Expedition 58 mission. The success of the abort and response to the investigation were reassuring when he willingly climbed on board the same spacecraft and rocket type less than two months later for his own mission. "This [abort] has made me feel even more confident about the way the Russians have designed the Soyuz spacecraft. It is very, very robust," he explained during an interview in Ottawa, Canada, in late October 2018, while the investigation was still happening.

Gerstenmaier had been there all along. He had started at the Lewis Research Center in 1977 before moving to Johnson Space Center (JSC) in 1980 to participate in the shuttle program. The transition from the research environment at Lewis to the fast-paced operational tempo at JSC was a challenge. "You know, you had to. prove yourself on console you had to get through certification. There were no training guides. There was no development stuff, no syllabus, you had to build all those training products for yourself. But it was a great time to come to NASA, it was really good to be there before the first shuttle flight and get a chance to get ready for that."[2]

He was thrilled to work with the amazing engineers from Apollo, and he worked directly with Gene Kranz. "There's a sense of excellence, right, that you're going to give 110%. It's important for you to describe what you know, and what you don't know. You have to be 100% transparent," he said describing the culture in the shuttle program. There wasn't the opportunity for formal mentorship, they were too busy getting ready to go fly the shuttle. Informally though there were many opportunities to learn: "You know when you work with him [Kranz] tough and competent was there. When you work with George [Abbey] it's exactly the same kind of attitude. They carried that excellence with them. So it was evident in day to day stuff."[3]

He sat in mission control during the first space shuttle mission way back in 1981. It was an exciting time; the shuttle was the launch system of the nation. He recalled, "We used to have a poster and it said, going to work in space. That's what the shuttle really was, a multi-user platform that could do stuff and build stuff in space."

He witnessed NASA's and Russia's first tentative steps together in space after the fall of the Soviet Union, as the partners ran joint missions on the Russian space station Mir. Tensions still existed between these former Cold War partners, especially as both agencies used to running their own separate programs found themselves working closely together for the first time since the Apollo-Soyuz mission in 1975. When Gerstenmaier became the Shuttle-Mir program operations manager, he immersed himself in the language and culture by living in Russia building trust with his Russian counterparts.

The Russians, Gerstenmaier recalled in a 1998 interview with NASA, impressed upon their American partners that ground communications with Mir were limited, that Russian communications must always come first, for safety reasons. He listened. He worked to the standard the Russians demanded. And, while being supportive of Lucid during her Mir mission, he always followed the Russian rules. And by following the rules, he earned their trust.

Flexibility

"If I told them I needed an extra minute of Shannon's time, I took one minute. I didn't take five," Gerstenmaier said. "If I got interrupted in the middle, I would give up the com. . . . So I just became totally accepted as part of the flight control team, so it was no different than being another Russian flight control team member in Moscow. So for me, that was very rewarding."

By the time he rose to his associate administrator position in 2004, he was unflappable. He was unsentimental. He was 50 years old, old enough to see quite a few space dreams broken. One of his first memories was seeing Sputnik fly over his rural house in Akron, Ohio. He dreamt of becoming a test pilot and attended the United States Naval Academy, but it was his time at Purdue University's school of aeronautics and astronautics that sparked his interest in space technology. Over the course of his NASA career, he became a calm presence on the ground, rather than the calm test pilot he had once imagined. Over the decades, NASA and the Russians depended on that calm voice.

Lessons

After 42 years with NASA, Gerstenmaier had managed a number of different projects, led major programs and helped guide the agency with its plans for space exploration. He had seen it all. He had worked with and learned from some of the best NASA leaders in the agency's history and was humble when sharing his insights on NASA's APPEL YouTube channel in 2018.[4] He discussed six lessons that in many ways reinforce those listed in the report of the Space Legacy group that was published two decades earlier.

- Lesson 1: Focus more on people to prevent mistakes and their consequences. This means "actually spending time with teams, interacting with the teams, so you really know what's going on with the program or project," he explained, rather than sitting in an office alone and looking at the data.

- Lesson 2: Reevaluate how NASA assesses the development of flagship missions. As missions move from paper designs to actual hardware, he said the key is to know management that works in one phase is "maybe not quite as effective in the other phase, so I think you need to adapt your management style and be aware that your project is changing . . . and look for different ways of doing business."

- Lesson 3: Realistically account for assembly, integration and testing. Making a new spacecraft is hard enough, he said, and unrealistic schedules (such as having the teams work nonstop for 24 hours a day for 400 days to make a deadline) will only lead to frustration. "Reality is, stuff is going to break. You do need some time in there to catch up, right?"

- Lesson 4: Recognize that first-time build challenges are inevitable. "The first time through is much, much longer. And you're talking about a learning curve, but that learning curve really comes down on the second and third times. You can really start honing those procedures. You know what fits, and what moves forward."

- Lesson 5: Establish a culture of transparent and open communication at all levels. "It's truly an art," he explained. "You want to put pressure on them, you want to hold folks accountable, you want to provide tough love, you want to push them to do new things, but also you don't want to be so hard on them that they are afraid to do anything, or they don't disclose to you when something is not going right."

- Lesson 6: Ensure oversight is organized and staffed appropriately at each phase of mission development. He added he tries to put people on a spacecraft team into a rocket team, and vice versa, as he's "trying to raise the level of expertise in program project management by letting these folks get to see different projects . . . it makes them carry extra work, but I think it's a really positive way of improving the overall knowledge and moving forward."

Investigation

Neither the Russians nor the Americans rushed things after that abort. The situation rattled a lot of people, as it was the first such incident in 35 years of launching Soyuz. Soyuz was also the only pathway for space station partners at the time. The shuttle was long retired and NASA's work with the private sector to develop new spacecraft was still underway.

NASA Administrator Jim Bridenstine was watching the launch at the Baikonur Cosmodrome in Kazakhstan and later commented, "NASA astronaut Nick Hague and Russian cosmonaut Aleksey Ovchinin are in good condition following today's aborted launch. I'm grateful that everyone is safe. A thorough investigation into the cause of the incident will be conducted."[5]

Gerstenmaier was heavily involved in the investigation and because of his relationship with the Russians they trusted him implicitly. He recalled, "All the failures that contributed . . . they were fully transparent with us, we got to see it all. I got to see the actual designs and get to see the actual hardware."[6]

"[The Russians]," explained NASA's Chad Rowe (NASA's director of human spaceflight programs in Russia), "had a really good idea about what had gone wrong that night, the night of the launch. They shared some of the information with us, and they said what their next steps were going to be. They went off and did this analysis to validate that their assumptions were good. Once they did that, they went through a large number of evaluations for how to mitigate the likelihood of recurrence in the future."

Although NASA wasn't invited to sit in on these meetings, two important things were going on. First, the Russians were sending constant updates to their major space station partner. Second, everybody knew NASA would have its own independent investigation simultaneously happening.

"While we're waiting, and this has been true of the major issues we've had with the cargo vehicles that haven't made it to orbit and other major issues, Kirk Shireman, my manager, the ISS program manager, will set up his own group of experts," Rowe explained. "We [will] take whatever we can learn from what we know, and then bring in our own experts and start to suppose what probably went wrong and . . . what we would do as a technical organization to mitigate recurrence."

Sharing

Eventually, Russia and the United States shared their results and found they had exactly the same conclusion — a deformed sensor in the rocket was at fault. Russia made some technical fixes, then ran a few cargo missions with the same rocket to make sure the fix was working. That process took a few weeks. By November, the rocket was certified once again for humans.

The timeline for sending a new crew into space accelerated, but everything worked out per the plan. Crew commander Oleg Kononenko was a veteran of several missions and easily passed the standard exams to direct operations on the Soyuz spacecraft. Rookie astronauts Saint-Jacques and Anne McClain shaved some time off their own schedules and also passed, with flying colors.

The partnership was confident in its work and preparation for the expedition 58 liftoff proceeded normally with attendees noting little difference from a normal space mission, even though the launch was three weeks early and followed a tricky abort. There were the same opportunities to speak with crew, the same filming opportunities before the launch and the same viewing perches nearby the launch pad.

In other words, almost everything went to plan. The crew members made it safely to space. The rocket, once again, worked marvelously. It was a triumph for NASA and the Russians. While Gerstenmaier kept his role quiet during those tricky weeks, his calm leadership behind the scenes set the tone for everyone's hard work solving the problem. He remembers, "He [general designer Sergei Romanov] probably understands how miraculous that abort was. That Nick could be back safe and we could turn around in six months and go fly him again. So, we were very, very, very lucky. But I don't think the outside world sees that. . . . These conservative engineers are always crying wolf. They're always saying it's hard. They're always saying it's difficult. We can have this abort, the crew is safe and everything is fine. They don't appreciate how difficult these abort scenarios are."

Gerstenmaier understands the risks associated with human spaceflight. He was there for Challenger and Columbia. He recalls Challenger, "was a big shock to the NASA system. These are really traumatic events. I had worked with many of the crew on Challenger. I knew them personally as friends. I went to church and school with some of them in the local area. And then to see them perish . . . it's really hard because you fail professionally, and you fail personally." Moving on is extremely difficult but necessary to honor the legacy of the crew and their commitment to exploring space.

There was an event that took place years later that helped put the tragedy in context. He said, "I was doing a Challenger memorial event for Mike Griffin down at the Kennedy Space Center, and

I was scheduled to speak on behalf of the shuttle program. June Scobee Rodgers spoke before me and I was totally blown away by what she said. She had no remorse for what we had done. She had lost her husband, but she took the tragedy and she turned it into the Challenger Learning Centers . . . to educate the next generation. I probably got more motivation from her speech than from anything I've seen, and, and that drives me even today."

Gerstenmaier's honesty and humility underlie his success as a leader in the high-stakes arena of human spaceflight. He is, "always aware we can have this really bad day. But if we've done everything to be prepared, and we've been totally honest with the crew, and they know the risk, and they're willing to take that risk, it's the right thing to go forward . . . I want to make sure that I've done everything I can. So if that bad day happens, I can look myself in the mirror, I can look at my actions and I can say, I did everything humanly possible to be ready to go fly."

The loss of Columbia rekindled those emotions: "It really hurt in a big way when we lost Columbia because I was sure that we would not, I would not let anything happen and but yet I couldn't control that [as manager of the ISS program]." He had a crew still in orbit that he had to take care of knowing that the Shuttle would be grounded for a period of time. He was able to channel all of his "negative grief energy" into something positive by taking care of the ISS crews. "It was not easy flying [those missions], [in] kind of an unknown period of time because we didn't know when the shuttle was going to come back. We had to make really hard decisions like dropping the crew size from three to two. I had to negotiate with the Russians about whether we even keep crews on orbit."

He had a very emotional telecon with his Russian counterpart Valery Ryumin. Ryumin was absolutely certain that the crews should be brought home. Ultimately, it was Genstenamier's reputation and the trust he had developed with Ryumin that kept the crew in space. During the teleconference, Gerstenmaier explained to Ryumin, "Valery, I know what you want. You want us to keep flying. And you think the way to do it is if we returned the crews, our government will then push hard to get the right resources and funds to get our crew back on orbit. But Valery, I think in my culture, if I returned a crew, there's not going to be any motivation for us to go back and fly again. It will actually have the wrong result and we will probably never fly again." Ryumin got very mad at that suggestion. "He cursed at me for five minutes in Russian. And in the end, he says, 'I believe what you said, Bill. We'll keep crew on orbit. We'll keep one Russian and one [American] and we'll figure out a way to make this work.'"

Relationships are important. This wasn't a situation where some manager at NASA is trying to negotiate a better position. Ryumin knew where Genstenmaier was coming from. "He [Ryumin] understood exactly where I was. He saw me live in Russia with him for a year, hand in hand, my heart was 100% for human spaceflight, just like his we had the same motivation. So that trust was so deep, that even though he didn't like what I told him, he knew what I was telling him was exactly the right thing."

For Gerstenamier, trust is an absolute requirement: "You've got to have 100% confidence and trust each other. If I'm having a bad day, I'm going to tell you I'm having a bad day. So, you make extra sure today. . . . You watch out for me today because today I don't know what's going on. But today I'm not quite as good as I usually am." Those are the checks and balances that make teams work.

The space station program is a great example of the benefits of collaboration. The other aspect of the program is diversity in that collaboration. As program manager, he found, "when you bring different cultures together and a different way of thinking and a different design structure, I think you end up with a much stronger, better program than you do when you just keep it insular within your own agency within your own activities. I think it's really important as we go forward for human spaceflight, that we continue that same kind of collaboration internationally."

"Bill was recognized by everybody as being technically well grounded and very astute. He was known to listen carefully, and to make his judgments based on good technical reasons," commented former space station program manager Wayne Hale when Gerstenmaier left NASA.

It was tough for those remaining who had worked with Gerstenmaier over the years. But his legacy and that of other retired NASA executives is still very much alive at the agency. Gerstenmaier always tried, "as a senior manager to question my own logic, [to consider] things that you think you really know, really probe those and make doggone sure you really know them and then spend a little extra time. If you don't understand something, and somebody explains it to you, it's okay to ask [more questions] and really gain that understanding. Humility is important because our business is so hard. That humility piece, it's got to be there. There are always unknowns. How do you tease those out? How do you understand them? That's the challenge we all have; how do you keep moving forward [when] you're worried about the unknown?"

In February 2020, Gerstenmaier became a consultant for SpaceX, just a few months before the California-based company prepared to launch its first human mission into space. At the time, *Ars Technica* said the move would be a boon for space station partners. "Few people in the global aerospace community have as much gravitas as Gerstenmaier, or as much understanding of how to build coalitions to explore space," the organization said in a story about the new appointment. "As SpaceX seeks partners — including NASA — to work with it on developing Starship to take humans to the Moon and Mars, Gerstenmaier is well positioned to offer advice, stitch together mission plans and open key doors."

Gerstenmaier's perspective: "I believe this [American] leadership role is absolutely critical to us. We cannot give this up. I think it's encouraging that the current administration is pushing very hard for the lunar stuff. . . . But we have to make sure that it's a sustainable program. The things we're trying to purposely put in place to build an avenue where the U.S. could be a leader, but yet still work with other nations and other partners that continue to keep going forward. You have to be eternally optimistic or you would have quit a long time ago because you're given these impossible jobs."

Leadership Insights

- Calm leadership behind the scenes sets the tone for the work solving the problem.
- Understanding why programs succeed helps achieve success. Understanding why failure occurs helps sustain success.
- Agile organizations are resilient, always evolving, and are willing to innovate and understand how to recover from failure.

CHAPTER 22

Searching for Solutions

"Coming together is a beginning. Keeping together
is progress. Working together is success."
— HENRY FORD

Most organizations have sentinel moments in their history that are part of the "urban myth" that is built into the stories told to new team members. There are many such moments that are part of NASA's history. From the "No handball allowed" sign that Alan Shepard put in Senator John Glenn's spacecraft Friendship 7 before liftoff to the "failure is not an option" mantra of Apollo 13, NASA is an organization built on the pillars of the past. Over time most great stories take on a life of their own; perhaps embellished in the telling, they become part of the culture of the organization. They are retold in the halls, after meetings, over a late-night beer shared with colleagues. In those stories are insights into what makes the organization work.

Responding to the challenge of a spacecraft that partially exploded halfway to the Moon, the problem solving by the Apollo 13 mission control team, engineers and astronauts is emblazoned in the five words "failure is not an option." There are few in NASA and the corporate world unable to recall those words from the film,

Apollo 13. For anyone in a critical team situation, it says it all. It takes little to imagine the tough and competent former marine pilot turned flight controller Gene Kranz assertively sharing that statement with the team. They worked together, creatively debating solutions. They took whatever the crew had aboard the spacecraft and figured out a way to create a new carbon dioxide removal system from paper checklists and duct tape. And, according to urban myth, the phrase "tiger team" was used to describe the process. Like many urban myths, that is not the full story.

Hypersonic

Walt Williams received his Bachelor of Science degree in aeronautical engineering from Louisiana State University in 1939 and soon joined the team at NASA Langley Research Center, where he worked on the stability and control of aircraft. Rising through the ranks he became the chief of the NASA High-Speed Flight Station in Muroc, California. The flight research programs challenged the teams with a myriad of complex questions associated with supersonic flight that were followed by later tests of hypersonic aircraft. While many people are familiar with the history of supersonic flight and the stories around breaking the sound barrier, even today the realm of hypersonic flight is often met with a quizzical "What's that?" It is essentially supersonic flight on steroids.

At hypersonic speeds, the molecules of air that surround the aircraft start to change by breaking apart or picking up electrical charge. This doesn't always happen at the same speed as is the case for supersonic flight. Instead, "hypersonic" refers to the point at which they start to meaningfully affect the mechanics of flight. Occurring at roughly five to ten times the speed of sound, from 3836 to 7673 mph, hypersonic flight creates an environment where small changes in one component can cause large changes in airflow around all other components. It is an environment intolerant of error.

The research program was pushing the limits of the flight test envelope, challenging engineers to use new materials such as titanium or cooled nickel with new designs that were highly integrated, instead of assembled from separate independently designed components. It also challenged the flight surgeons, the medical doctors, entrusted with the health of the test pilots. As in the case of the engineers, the doctors were exploring uncharted territory and Williams wryly recalled, "And [they] had a discussion with one of the doctors and Joe Walker [X-15 pilot]. And this doctor asked Walker whether he had fainted or not during the X-15 flight. And Walker said, 'My God, no!' 'Well, how do you know you didn't?' He said, 'Because everything I did one second depended on what I did the last second and I'm sitting here talking to you.'"

Working together in interdisciplinary teams, they were able to overcome these challenges, ultimately leading to the design of the North American X-15 rocket-powered aircraft, and lessons they learned would later be incorporated into the design of the space shuttle, enabling it to fly 25 times the speed of sound or five miles a second.

Tiger Teams

Williams's teams were made up of technical experts who found solutions through discourse, debate and occasional conflict, with resolution found in a shared solution that would be readily tested in subsequent flights. The data-driven **"test, learn, try again"** process worked and proved to be a successful approach to developing complex technical projects. Williams brought the same approach to NASA, where he directed the worldwide tracking network and worked as the flight operations director in the Mercury program, rising through the ranks again to become the deputy associate administrator in the Office of Manned Space Flight and NASA HQ and ultimately the NASA chief engineer.

The origin of NASA tiger teams can be found in a publication he wrote in 1964, years before the Apollo 13 mission, that described the principles of project management associated with the design of complex technology. The authors used the phrase "tiger team" to describe using a small group of technical experts that have to solve a specific problem, generally in a short time frame. Defined as "a team of undomesticated and uninhibited technical specialists, selected for their experience, energy and imagination, and assigned to track down relentlessly every possible source of failure in a spacecraft subsystem or simulation," tiger teams became part of the organizational culture in the early days of human spaceflight.

Undoubtedly, selecting individuals for their technical experience and energy makes sense. The interesting elements of the definition include the descriptors undomesticated, uninhibited, energy and imagination. One can imagine the chaos of an environment where each team member energetically, without inhibitions, shares their best ideas with their peers. With voices raised, chalk flying, wielding slide rulers, technical notes and diagrams waved about on yellow ruled writing pads. From chaos came consensus, problem resolution and success.

George Abbey was struck by the unique approach used by the NASA teams. "When I first came to NASA, I'd been accustomed to working in the Air Force in research and development activities [in the joint Air Force, NASA Dyna-Soar program]. At NASA it was very apparent that there was a great cultural difference between what I had experienced before [in the Air Force] and [what I experienced] at NASA." Creating a platform for candid communication was one of the reasons why tiger teams worked.

Bruce Tuckman was a professor of educational psychology at Ohio State University around the time Williams published his paper describing tiger teams. He described what has become widely accepted as the various stages of group development. "Teams are formed, this is followed by the storming phase as they get to know one another, figure out what they're doing, how they're doing it and develop trust in each other." Tuckman felt it was important to resolve disagreements and personality clashes to progress to the next stage, norming, where the team begins to work together to then arrive at the performing stage where they are able to productively work together. Tuckman acknowledged that dissenting views are expected and allowed as long as they are expressed in a manner acceptable to the team. Later Tuckman added a fifth stage called adjourning that was associated with completing the objectives of the team. Others have referred to that stage as transforming, with a focus on the impact of implementing change in an organization.

Switching from the group to the individual perspective, within many organizations when people are assigned to a special team, they may ask, "Why am I here? What are we supposed to do? Why do I have to work with these people? When will I finish?" Some top performers who get repeatedly asked to be on special teams internally wonder, "Couldn't they have asked someone else?" Those questions are not asked by members of a tiger team. They know why they and the other team members are on the team and what the goal of the team is. Working in a competency-based culture, there's no question that each team member is there to share their experience. There is a strong, shared commitment within a NASA tiger team to get down to business and solve the problem.

Perhaps in a culture that values openness, dialogue, brainstorming and debate as techniques to achieve excellence, it is not surprising that many at NASA feel tiger teams go directly from forming to performing. Time is an organization's greatest asset and there is no need to storm and norm if the culture of the organization is taking on and resolving significant challenges. In the high-stakes early days of NASA, team members were different from today's tiger teams but nonetheless they were effective. The sometimes-heated disagreements were simply part of the path to consensus and resolution.

Tiger teams should not replace the normal project management structure within organizations and an understanding of when to use them is important. In general, tiger teams are particularly suited to time-critical problems that require solutions within days to weeks, not months or years. The task should be focused on a particular problem and the individuals assigned to the team should be experts in that area. Clarity of purpose and a defined timeline help avoid scope creep that could lead to findings or recommendations that are overly broad and deal with the problem at a high level but may not actually provide a solution. If possible, team members should be given sufficient time within their existing schedule to participate in team activities and the team should be appropriately resourced.

The Apollo 13 tiger team is an excellent example of rapidly mobilizing a team to bring the crew safely back to Earth. There were many different subtasks that needed to be solved by different groups of experts. Kranz recalls responding in the early hours after the explosion. "The key thing here was I didn't form the tiger team — the [foundation for the] tiger team to a great extent always existed. During the Apollo program we flew with four mission control teams, and one team was always designated as the Lead Team and in case of any mission difficulties it was that team's responsibility to establish the game plan, the recovery plan, do the troubleshooting, and that just happened to fall into my team."

Kranz and his White Team continued their shift for an extra hour before handing over to Glynn Lunney and the Black Team. Kranz then put together a tiger team to concentrate on the engineering and procedures needed to get the crew home. "We had many problems here — we had a variety of survival problems, we had electrical management, water management, and we had to figure out how to navigate because the stars were occluded by the debris cloud surrounding the spacecraft. . . . We were literally working outside the design and test boundaries of the spacecraft, so we had to invent everything as we went along. There were many lead people, but the three people I would name were Arnie Aldrich who handled the checklist, John Aaron with the power management, and Bill Peters who looked at the lunar module and tried to figure out how it could be most effective as a lifeboat." Within 24 hours, they had developed the solutions they would use to bring the crew home.

Shoulder Injuries

> "I didn't want to just know the names of things. I remember really wanting to know how it all worked." — ELIZABETH BLACKBURN

Unlike Apollo 13 where the team had to find immediate solutions, the NASA spacewalk shoulder injury tiger team used a longer timeline of five months to develop recommendations and mitigation strategies. Over the preceding three to four years there had been growing concerns about the frequency and severity of shoulder injuries experienced by astronauts during their underwater training sessions. The cause of the shoulder injuries was unclear, and the approaching wall of spacewalks required to complete space station construction brought a degree of urgency to resolving the issue. It was clear to the mission operations team that losing an assigned crew member from a shoulder injury within a few months of launch could have significant mission impact and affect the sequence of the space station construction missions.

The team's mandate was clear: find the cause and develop a solution. Experts from flight crew operations, space and life sciences, mission operations, engineering and spacewalking were assigned to the team and a number of specialized consultants in orthopedics and sports medicine were used as needed. Working inside a spacesuit is a unique challenge, with the suit constraining joint movements somewhat like an exoskeleton. NASA engineers had designed two different types of shoulder joints for the space-suit, one of which pivoted with the movement of the shoulder and the other with a fixed opening. Many astronauts preferred the pivoted design or the so-called pivoted hard upper torso (HUT) for its increased shoulder mobility. Unfortunately, there had been a catastrophic failure of a pivoted HUT shoulder joint during training and the suit had been taken out of service. The planar or fixed-opening version of the HUT was the only option for training and use in space.

To understand the possible cause of the shoulder injuries, the team would first have to understand the normal biomechanics of shoulder motion, the biomechanics of shoulder motion within a spacesuit, the anthropometrics associated with the sizing of the suit as well as the loads and forces placed upon the shoulder during suited training.

It was a daunting challenge. They used laser anthropometric studies to determine shoulder joint mobility in suited and unsuited crew, and a load-sensing handrail attached to an underwater platform to understand the forces associated with different body positions used by crew during training. The data found that the planar HUT significantly decreased the range of motion of the shoulder joint while certain body positions, particularly working inverted during training, further compromised shoulder joint mobility. The restricted mobility during suited training resulted in impingement of one of the shoulder joint muscles called the rotator cuff that resulted in either partial or complete tears of the muscle which, in some cases, required surgical repair.

With that knowledge, the team made immediate action rec-ommendations to use a shoulder harness in the suit that reduced the impingement while also reducing or, if possible, eliminating inverted suited training. The success of the tiger team was related to the clarity of the objective, the wide range of technical resources and expertise available to the team and the support of management to reduce the workload of the team members to allow them to participate in the team's activities.

Elements of Success

The use of tiger teams has spread beyond the realm of spaceflight. In fact, many recall the use of tiger teams by different branches of the military, in some cases before the Williams publication. Perhaps it was the influence of Williams's coauthor and Navy test pilot Scott Crossfield, who would later become the chief test pilot for North American Aviation working on the hypersonic X-15 aircraft. It is not the origins of the tiger team concept that is relevant, it is the opportunity to selectively add the tiger team approach to project management in different organizations.

Companies involved in innovation to develop new technologies or IT companies investigating software anomalies or security-related incidents have all benefited from the tiger team approach. In some cases, sub-teams are assigned to specific tasks that will help the overarching tiger team find a solution to the problem. Overall the common elements of tiger teams that are associated with success are a culture that supports speaking up, one that is dedicated to building competency and expertise, a clearly defined deliverable and timeline with the allocation of a small group of experts given the appropriate resources. Perhaps the need for experts who are also undomesticated is less important today but Williams's assertion that experience, energy, imagination and a relentless commitment to find "every possible source of failure" is as important today as it was at the beginning of the Mercury program.

Leadership Insights

- Immediate problems generally require immediate solutions that can often be found by technical experts working together, but only if they are given the opportunity.
- Test, learn, try again — repeat.
- Candidly sharing ideas, planning and conducting tests of those ideas then succinctly communicating recommendations are the main reasons why tiger teams work.

CHAPTER 23

The NASA Way

"Leadership and learning are
indispensable to each other."
— JOHN F. KENNEDY

Thousands of visitors, contractors and employees drive through the main gate at Johnson Space Center every month. Turning from the appropriately named Saturn Lane onto the Center's Second Street, the main security gate is the first stop. No one gets on site without a badge. When granted access, everyone will get to drive by the rocket park. It is impossible to miss. The Little Joe II, Mercury-Redstone and the Gemini-Titan rockets point skyward as though ready to launch, each a reminder of the early days of the space program. The largest, the Saturn V, is now housed indoors in a low-lying building to protect it from the elements. For years, the 30-story rocket, on loan from the Smithsonian's National Air and Space Museum collection, lay on its side unprotected as though waiting to be positioned vertically for launch. Towering 363 feet above ground, the Saturn V weighed 6.2 million pounds after it was fueled on the launchpad. Liftoff was a ground-shaking controlled explosion of 7.6 million pounds of thrust that took the Apollo astronauts on their journey to the Moon. It is impossible to drive by without looking at the rockets and thinking about the astronauts who rode on top of them and the teams of people that sent them to space. The view is unforgettable. For most, the feeling is a mixture of awe, inspiration and incredulity at the sheer magnitude of the achievement and the technology. For others, it is a symbol of NASA's culture.

Organizational culture has been studied by academics and business leaders for 70 years. There are many definitions, descriptions and attributes associated with different corporate cultures. Some are complex and some relatively simple. A definition that is relatable to most describes corporate culture as the way people work together in an organization. It is the way things get done. It can be a source of pride, learning and growth or it can be a dysfunctional, demotivating source of job frustration. At its best, the NASA culture is one where individuals are committed to learning and building their competency, where they are proud to be the best team members they can be to support the exploration of space. Some would describe the NASA culture as vibrant, dynamic, and ever-present. It's a place where the best of the best work together to make the impossible possible. However, these phrases describe the culture, they don't define the culture. A recent conversation with former NASA Administrator Mike Griffin provided the best definition of the NASA culture: "We're an engineering culture. And that's what we should focus on."[1]

In clarifying the engineering culture, he paused and thoughtfully responded, "When I say an engineering culture, what I mean is while people are dedicated to learning, a good engineering culture is one that settles arguments with a facts-based discussion. They're always trying to improve; they're always trying to make it better. It's about what you can do, technically, to be state of the art. And of course, what we do at NASA is very, very, very hard. And when it goes wrong, people can die. But the best of NASA, to me, is the best of engineering culture."

Candid fact-based discussions informed by the best available data were as important for success in the early days of Mercury and Gemini. It was critical when George Low led the return to flight after the Apollo 1 fire, Richard Truly led the return to flight after the loss of Challenger and Mike Griffin led the return to flight after the loss of Columbia. "When we're at our best, we're willing to disagree. Because we don't see the facts the same way or, more often, we don't put the same weighting factor on a given fact. We may agree on the fact, but one person will think it's more important than another one, and the other will think it's less important. And those are the kinds of discussions and arguments that should surface and do surface when we're at our best. It's being willing to listen to the argument, no matter from whom it comes. It could be the youngest person in the room who has the correct slant on things. Or it could be the oldest person in the room that everybody else thinks is outdated. But the argument is the argument, and it should be discussed on its merits. And when we're at our best, that's what we do."[2] A hard-learned aphorism from NASA's top leaders over the decades would be, "Don't tell me what you think I want to hear, tell me what I need to hear and why I need to hear it."

In most organizations it takes courage to speak up. There were times when that was the case at NASA as well. It takes strong leadership to mentor others in the art of discussion where team members can passionately disagree with one another during a meeting and then go out for a beer together afterwards. The early NASA leaders understood the merits and techniques of candid discussion and successfully created a culture where speaking up was welcomed. Former director of Johnson Space Center General Jefferson Howell — call sign "Beak" — recalled, "If you don't keep score and people don't take it personally, it's an excellent way [to resolve issues]. I think Gene Kranz and the culture he set up in mission control was very healthy. It's carried on to the current mission control team, the way they're willing to argue with one another and then resolve it and then get on with it and pat each other on the back say let's go."[3] Remember though, it doesn't work if "people take things personally, hold grudges and [want to] get back at each other. It's very typical for any large organization that you're going to have different segments with their different ways of doing things."

Perhaps it was due to his credibility as an astronaut combined with his expertise as an engineer that led John Young to become one of the most respected individuals in human spaceflight. A veteran of Gemini, Apollo and the Space Shuttle programs, he had walked on the moon and commanded STS-1, the first flight of the space shuttle. Whether it was at a Monday morning astronaut meeting, or a briefing with agency leaders, John always said what he thought. Mike Griffin recalls, "John spoke up no matter what." In those moments where the culture wasn't at its best, "there were occasions when he was punished for that. He finished his career as a senior staff adviser to the director of Johnson Space Center. That wasn't where he really should have been. But it didn't ever prevent John from speaking up. I would say John Young exemplified exactly what I'm talking about, even when it cost him."[4] Courage, integrity, commitment to doing the right thing and speaking up are all values embodied by NASA leaders that are critical to success in the high-stakes arena of human spaceflight. They are not easy.

In all organizations, there are times when the values resonate and there are others where they seem to have been forgotten. During the return to flight activities after the loss of Columbia, Griffin focused on getting people to speak up when they disagreed, on getting teams and leaders willing to listen. "That doesn't mean that you get to repeat yourself ad infinitum. The ability to put your argument on the table is not the same as continuing your argument until you've beaten people down. It has to stand on its merits. There's a happy medium between listening and listener fatigue. But when we're at our best, we welcome people's disagreement, and we settle it the best way that we can; when we're not at our best, we punish people by reassigning them because we don't like what they have to say."

If speaking up were easy, there would be no issue. Unfortunately, it is hard and because it's hard often it doesn't happen even when requested by leaders. **Speaking up thrives as a corporate value in high-trust organizations where individuals don't have to worry about what they say.** In low-trust organizations it requires courage and occasionally a willingness to accept whatever the consequences of speaking up may be. Leaders in government agencies must be willing to weather the storm of political feedback with integrity and truth. Doing what's right may not be expedient. But in some cases, others are able to recognize what is right. The challenge for leaders is succinctly communicating the issue and the underlying data that supports their recommendation.

Changing organizational cultures can be difficult. When former Director of Goddard Space Flight Center and NASA Associate Administrator for Space Flight Joe Rothenberg spoke with the team at Goddard about non-technical systems engineering challenges, he commented, "people are the key ingredient." People create, modify and sustain the culture in organizations and Rothenberg believes, "you can't change the culture by changing the culture. You've got to change the value system, what they want to do, their motivations, and then the culture will change."

Strategic planning, workforce surveys to assess corporate values and engaging the workforce in a new corporate vision are tools that can help change the way things are done in an organization. Rothenberg mentioned, "you have to create a vision. Roy Bridges [former director of Kennedy Space Center] was really good at this. It's got to be a pull, not a push. A push is a really tough thing. But if you can get a pull and attract the cream of the crop, that's what you want to do."

When Bridges had to change the size of the workforce in the shuttle processing program, he wanted to create something new that would set the stage for the future of human spaceflight. "His notion was to make something they want to go to versus something to just push them off the program and that's when he established his idea of a commercial spaceport,"[5] Rothenberg said. He describes Bridges as a "thought leader" who was engaged with his team and worked hard to implement a new vision. It changed the culture at Kennedy (Space Center) and helped prepare the center for an era of private sector launches.

Engage all of the employees; the whole team is important. When Rothenberg started changing the culture at Goddard one of his team, Kathy Nado, said, "He went through a really disciplined process of defining what was going on within the center then put together a team of next generation leaders to define where it should be." Nado had suggested to him, "You have to skip down a level, [ask] the next generation — the folks that are division directors or smaller program managers to define what the organization should look like."[6] Rothenberg, always amenable to anyone with a good idea, decided to "hold focus sessions at lunchtime. We'd have pizzas for 500 people. . . . And people would come and then we'd tell them what we're thinking about long before we had made a decision. We held [additional] focus groups and then had this strategic planning group [that could] break up to hold focus groups whenever they wanted." His approach worked. The engagement helped. He "established these employee teams to get them feeling that they were part of the change. We listened to them. And they saw their ideas reflected. And if they didn't get reflected, they understood why."[7]

Former astronaut and director of NASA's Glenn Research Center Janet Kavandi used a similar approach to engage her team in incorporating expected behaviors into their culture. She created a poster outlining the desired behaviors and values based on the center's namesake, Senator John Glenn. The first letters of each behaviour spelt the word "HEROICS." Their shared commitment,

"Helping All to Succeed, Excellence, Respect, Openness, Integrity, Cooperation and Safety," resonated with employees at the Center, and by documenting and publicizing these expected behaviours, she succeeded in strengthening the culture.[8]

Ed Schein, Emeritus Professor at the MIT Sloan School of Management, is a global expert on organizational culture who feels that understanding corporate culture is a much more complex undertaking than attributing it to "the way things are done" in an organization.[9] He has described three elements of culture: artifacts, espoused values and basic organizational assumptions. Artifacts are visible indicators evident in and characteristic of the work environment. At NASA, visible indicators are everywhere and reflect the profound commitment of the workforce and their pride in what NASA has accomplished. When Dan Goldin got rid of the "worm" logo in favor of the original "meatball" logo, the unspoken message was a return to the Apollo-era culture and its commitment to competency, controlling risk and data-driven decision making. The NASA meatball had evolved from a logo to a brand, a brand defined by pride in operational excellence.

How organizations respond to adversity is a great example of the strength of corporate culture. After three tragic losses, NASA recovered through building competency, controlling risk, candid discussions and data-driven decisions. The response of the team at Kennedy Space Center to the cancellation of the Space Shuttle program was equally powerful and memorable. With a camera placed on the roof of the 525-foot vehicle assembly building used to prepare the Apollo Saturn V rockets and the space shuttle for launch, the team cleared the parking lot and filmed a video in which hundreds of team members came together to form the outline of the space shuttle. It is a compelling, bittersweet reminder of a remarkable era of space exploration. They were not responding to a management request; they were expressing their pride in being part of a program that had become an international icon.

Pride can be either a help or hindrance in organizations. Leadership expert John Maxwell has described pride as a leader's greatest problem — perhaps the antithesis to humility.[10] Maxwell points out that individual pride can adversely impact teamwork, learning, building relationships, and it may hinder individuals from reaching their full potential. That is not true of team pride. Team pride gets you to the Moon, it gets you to the Super Bowl, it wins you the Stanley Cup. Many highly effective NASA leaders are very humble and while some may fall victim to the problems of personal pride, most are very effective at building team capacity and pride. If people are not proud to be part of an organization, it is likely they can't relate to the corporate values and don't feel good about working together; it is hard to imagine success under such circumstances.

Team pride permeates NASA. The unspoken dress code at NASA is a reflection of its history and the pride in being part of a peak-performing team. Business attire is required to work in mission control and the launch control center. Logo attire is worn by members of the workforce for day-to-day activities; the NASA logo is prominently displayed above lettering that portrays an association with a mission or part of the organization. In a short walk around Johnson Space Center a sharp-eyed observer will note the ubiquitous polo shirts with the increment identifier or the phrase "commercial crew" worn by the astronauts, or the blue flight suits they use to fly in the T-38 aircraft.

There are still shuttle-era mission shirts inscribed with "STS-" followed by a mission number worn by management astronauts, training teams and members of the mission control team. The shirts have become a living library of NASA acronyms: MOD, FCOD, SLSD, Flight Medicine. Many of the shirts are purchased from Land's End, where there are hundreds of NASA-specific logos on file. The visibility of the logo throughout the workforce is an excellent indicator of the morale and pride in being part of the NASA team. Similarly, there are plaques on the wall for each mission overseen by mission control, a ceremony to place the plaque on the wall, photographs from missions adorning the halls. Visiting any NASA center, it is impossible not to feel part of the agency's incredible history and pride in its accomplishments. Perhaps that explains the current popularity of NASA-brand attire; it enables everyone to feel a little bit of pride in what NASA has done.

Schein described these visible indicators in a presentation at Google about corporate culture. "When we arrived here, we saw what the anthropologists would call the artifacts, or the artifices, the creations of culture. [These include] the buildings, the way people dress, the structures, how you get into the place, all that stuff is the visible, feelable, smellable, hearable part of culture, which is real, but is just the surface manifestation." The deeper layer of corporate culture includes the espoused values: "'We're creative, we're fast, we're team-based' — those words are espoused values. They are not necessarily what is at even the deeper level that would explain in detail day-to-day behavior. I think of that [espoused values] as your taken-for-granted assumptions. It's sort of the automatic way you learn to behave around here."[11]

The espoused values at NASA are a complex interwoven blend of occupational cultures with corporate culture. The NASA workforce includes engineers from many different areas of specialization, scientists, physicians, pilots, aerospace experts, management staff and support personnel. A typical NASA team member has at least one degree, with many having multiple postgraduate degrees with additional work experience. The agency attracts the best and the brightest, so it is no surprise that knowledge and learning are de facto espoused values. The many professional occupations represented within NASA create an organization of blended occupational cultures. It is true that NASA is primarily an engineering culture; more importantly, it is an engineering culture that has been impacted by the occupational cultures of other NASA professionals. Perhaps the best way to view the espoused values of such a blended professional culture is to think of the various differences as an overlapping Venn diagram and look for the common elements. These would include passion, commitment, strong work ethic, evidence-based or data-informed decision making, knowledge and ability to learn.

The deeper NASA culture would be described by Schein as the underlying assumptions that "really drive behavior." Those assumptions may be inherently obvious or more subtle and they reflect what has worked in NASA's history: tough, competent, speaking up, doing the right thing, truth, integrity. They are learned, lived through peer-to-peer communication, leadership and followership. They are critically dependent upon how leaders use power, positional authority and influence. Despite the learned value of speaking up, the first time a leader ridicules, or directly or indirectly punishes someone for speaking up, information flow dries up. **If dialogue and debate are permitted to become personal, the incentive to share ideas disappears.** Corporate situational awareness is critical for leaders. Just as the astronauts, mission control teams and pilots have to maintain operational situational awareness, leaders must understand what is really happening in the organization to be effective. One of the NASA leaders with the best corporate situational awareness was George Abbey.

> "The important thing is not to stop questioning."
> — ALBERT EINSTEIN

Long before experts would speak of the importance of mindfulness and relationship building in organizations, Abbey was building relationships within NASA and the contractor workforce to provide George Low with insights into what was really happening in the development of the Apollo spacecraft. Low innately understood that to succeed he needed information on the problems that he wasn't hearing about through the regular management reporting process. That corporate situational awareness was critical to the success of the Space Council in the early days of the space station program. "I think it was with George [Abbey]'s joining the Space Council that we really were able to communicate effectively and . . . to understand from NASA's perspective their requirements and practicalities that really allowed this [the Phase-1 Shuttle-Mir program] to move forward," said Albrecht.

Jim Wetherbee recalled working with Abbey. "When you're working in a large socio-technical organization like NASA, the best way you as a leader can help your workforce do better is to understand the culture in which they are operating. And how to inspire them. You've got to understand the spaceflight culture. What George did was the way you are supposed to change the culture."[12] Abbey cared for his team while at the same time doing the right thing for the program. He combined chili cook-offs with safety days and a longhorn cattle project with community outreach to engage students' interest in science and engineering.

Abbey's underlying assumptions permeated the culture at Johnson Space Center at the time — commitment, integrity, competency, respect, punctuality, doing the right thing and giving opportunities to those with talent. He was probably the best judge of talent in NASA's history and was willing to give people opportunities to become agency leaders. Mike Griffin described him as "an extraordinarily good judge [of people]. He could pick good people. Whatever his faults and good points, George's strongest good point was being able to judge good people. And everyone knew he would hold them to account."

Bill Parsons, former director of Kennedy Space Center, worked closely with Abbey as the deputy director of the Johnson Space Center. He recalls, "This is the George Abbey influence, and it probably comes from the George Low influence. That is, you pick somebody and then you mentor them and if they have what it takes, you put them in various leadership positions, and you keep moving them up if they continue to be successful. Working with George Abbey and many, many other leaders within NASA, as I got to know them, it helped me move up. They gave me increasingly more responsibilities along the way. If I was successful, I got another opportunity to do something even more impactful. My whole career was really a result of great mentorship and then being given opportunities, performing at a certain level and learning as I went along. I didn't realize it when he offered it to me and I surely didn't know that I'd been building on my skills and on my experiences . . . but believe it or not, I was ready. But it didn't feel that way when [I was] given that opportunity. The NASA that I knew was the George Abbey NASA, because George influenced human spaceflight so tremendously during his time. It carried over into me because Jay Honeycutt worked with George Abbey — that was a **culture of mentorship, learning and giving talented people increasingly more responsibility."**

> "Destiny is not a matter of chance. It is a matter of choice. It's not a thing to be waited for — it is a thing to be achieved."
> — WILLIAM JENNINGS BRYAN

Abbey's approach was not appreciated by everyone; many felt that it was based on doing what Abbey wanted. Some were overtly critical and used the phrase "friends of George," implying that promotion was based on favoritism. His probing questions during meetings and his commitment to holding people accountable undoubtedly made the ill-prepared uncomfortable. Abbey's questions were probing but never unfair.

Wetherbee recalled: "I was the director of flight crew operations and he would ask: What went on last week? What's coming up next week? What did you learn? What issues do you have? What challenges do you have? What, are you working on? And I would have to answer all of the questions. He was getting me to open up the books and explain what was going on in FCOD [the Flight Crew Operations Directorate]. That's a process of accountability."[13]

Wetherbee learned you didn't always have to have the answers, you had to be truthful and have a plan to work on whatever problem you were facing. If a particular issue was not disclosed, Abbey would ask about it anyway if it was important. He always knew what was going on, never forgot anything and was relentless in his commitment to safety and doing the right thing. Some liked that approach and thrived, others didn't and complained. No one can argue about what he achieved. Abbey built a culture of safety, competency and accountability, mentorship and learning that was similar to the one George Low had built in the Apollo era. There was never a mishap under Abbey's leadership as everyone knew that launches would take place when it was safe to launch, not necessarily when they were scheduled to launch. Issues were identified, worked, resolved and the highest flight rate in NASA's history occurred without incident.

Mike Griffin became NASA administrator after the loss of Columbia and had to assess the underlying assumptions in the workforce that might have contributed to the loss of the vehicle. The values of the leader rightly or wrongly become quickly recognized and for the most part incorporated into the deepest level of the corporate culture during their tenure. Given the candor of his interactions with the NASA team, the press and the administration, his values were readily understood. These were critical in successfully returning to flight. His fearless integrity, insistence on data-informed decision making, discourse, debate, his thoughtful analysis and search for critically missing pieces of information to make the best possible decisions and his willingness to listen to dissenting opinions all shaped the way in which the agency worked.

There was a fundamental shift in the underlying assumptions of the agency's culture. He pushed back within NASA and within the administration to achieve the results that were critical to success and without his leadership it is doubtful that NASA would have returned to flying the shuttle and finishing the construction of the International Space Station.

The story of NASA's culture is one of its leaders. The espoused values and underlying assumptions that shaped the culture changed over time with shifting emphasis to one approach or another. It is a study in the legacy of its leaders. It has been shaped and influenced by the many moments where it achieved the impossible. That's the NASA way.

Leadership Insights

- If an organization universally recognized for its achievements can normalize deviant behavior, all organizations must be vigilant to the risk of culture creep.
- Corporate situational awareness should include understanding corporate culture.
- Leaders that live the culture can change a culture.

AFTERWORD

Dave Williams

There is no lack of material for those interested in studying leadership. Bookstore business sections have shelves of material dedicated to the subject, ranging from the leadership styles of the military to leadership lessons based on fables. With so much material, why did we set out to write yet another book on the subject? In part, the answer is related to personal experiences working with many of the individuals interviewed in this book and watching the success associated with their leadership styles. Perhaps the answer also lies in an attempt to underscore the importance of continuous learning from personal experiences and the stories of other leaders to help each of us to build our repertoire of leadership styles and grow as leaders.

The perspectives on leadership at NASA we have shared, simply put, work. They have been validated through the results that were achieved with their widespread utilization over the past 60 years of human spaceflight. They are important for both aspiring and experienced leaders in any sector. And, like many aspects of leadership, will be challenged, criticized, analyzed and in some cases adopted. The one common element of leadership is everyone has an opinion, yet few recognize the unique relationship between the leader, the senior team and the broader organization that is so critical to success.

Leadership expert and author John Maxwell has characterized leadership with one word: influence. Many say that leadership is visionary, inspiring, optimistic, should be focused on providing resources or enabling others to succeed. There are long complex definitions and descriptions of different styles of leadership, but the best is often the simplest: influence. Leadership, the art of influencing others. There is elegance in the simplicity of the definition that reso- nates with styles ranging from the quiet questioning of George Low to what some describe as the frenetic style of Dan Goldin. From the "failure is not an option" approach of Gene Kranz to the reas- suring questioning of the mission control team by Glynn Lunney in recovering from the explosion on the Apollo 13 spacecraft, to the integrity and commitment to data-driven decision making of Mike Griffin in returning the shuttle to flight after the loss of Columbia, different moments need different approaches to influencing teams to achieve their best.

There are essentially three different outcomes when leaders influ- ence others. Leaders may have a minimal impact, negative impact or positive impact on an organization. Those leaders committed to achieving organizational excellence are typically introspective, continuously asking themselves if the approach they are taking is working. The introspective leader quickly realizes that the effect they are having on the organization can be either positive, neutral or negative. Large organizations have an element of inertia and tend to resist change, just as a large rolling ball exhibits the property of inertia described by Newton in his first law of motion. Objects and people tend to "keep on doing what they're doing" unless acted upon by an external force. Some leaders either don't aspire to or never find an approach to change the organization they are leading, and things continue as they always have. In some cases that might be appropriate in the short term, but over time there are few organizations that are not required to change and adapt, to approach new challenges with innovation and agility. When a leader is able to provide positive influence working with a team, problems are solved, challenges are overcome and organizations grow.

Positive influence is not always popular. Change is difficult and despite many within the organization knowing what is needed, there are always those who resist doing what needs to be done. Dan Goldin had to deal with more than a few longstanding career civil servants who fell back to using the "we can wait him out" approach, with many thinking "this too shall pass." Some were more Machiavellian in their approach, actively trying to bring about his departure to protect the stability they valued, the inertia that had been created.

Mike Griffin maintained his integrity in doing what was right for NASA to safely return the shuttle to flight despite possible risk to his career. George Abbey has been criticized by many, but those who worked with him witnessed and understood his relentless commitment to doing the right thing. While transformational leadership sounds exciting, it is not for the faint of heart. Personal resilience in the face of overt, and more often covert, criticism is critical. Vision, commitment and integrity in doing what needs to be done are all easy to write, even easier to read and often extremely difficult to enact.

Introspection can also help leaders determine if they are having a negative influence on an organization. Both positive and negative change can be met with pushback from the organization and pushback alone doesn't help leaders understand if they are going in the right or wrong direction. Recognizing that bad leaders, and occasionally even good leaders, can take an organization in the wrong direction, the best leaders create time for a solo daily debrief to thoughtfully assess how things are going. Reflecting on the opinions of trusted colleagues, careful consideration of objective performance metrics and a simple willingness to consider the possibility that things may be going in the wrong direction are all important in ensuring ultimate success.

It is also worth considering the process of organizational influence. Modern leadership is moving away from the idea of a single heroic leader who guides the organization towards the concept that optimum results are achieved through leadership, followership, teamwork and a sustained commitment to competency. Every organization has a hierarchical chart of its management structure and it is frequently one of the first few slides in a corporate presentation. Let's forget for a moment the merits of leaving the org. chart in or taking it out of a pitch deck. Given that most large organizations are hierarchical, it is appropriate to focus on the nature of influence in those organizations.

Influence follows the flow of communication. It can occur in a downward direction, what most would call traditional leadership. It can occur horizontally through peer-to-peer influence and, in some organizations, it can occur in an upward direction through a process called followership. NASA has effectively used followership throughout its history from the early days of the chaotic, often heated, team meetings described by George Abbey to the current era, where leadership and followership training with the National Outdoor Leadership School (NOLS) are included in the expeditionary behavior training for long-duration astronauts.

Oscar Wilde said that "imitation is the sincerest form of flattery that mediocrity can pay to greatness." The quote is frequently shortened, the word "sincerest" has often been replaced by "greatest" or "highest" and, notwithstanding the temptation to get sidetracked into a discussion about plagiarism or mediocrity, one interpretation suggests that the essence of the quote is the assertion that what works gets repeated.

Incorporating a core value of speaking up into corporate culture works. It helps high-stakes organizations succeed, but only if leaders listen. **Tied to speaking up is the idea of listening up, for leaders to thoughtfully assess the information that they have been given when making their decision on how to proceed.** Despite the many personality differences of the NASA leaders, a common attribute associated with their success was a willingness to listen, to encourage discussion and debate in the analysis of best available data to make the best possible decision. In a high-stakes operational environment where readiness is critical, misinterpreting the data or downplaying what the data is saying in favor of meeting schedule pressures or other corporate objectives can be perilous.

Those who have worked closely with George Abbey understand his philosophy of launching when you are ready to launch, not necessarily when you are scheduled to launch. There are few NASA leaders who witnessed the losses of Apollo 1, Challenger and Columbia. He is one of a very small group, and the lesson that NASA learned through tragedy is as important for the future of human spaceflight as it is today. The process of determining readiness is relevant in many different sectors and whether it is embarking on a new clinical procedure, drug or vaccine or bringing to market a new piece of technology, ensuring readiness is critical. There are many regulatory agencies in different sectors that try and answer that question and many organizations would benefit from building a process to determine readiness into their corporate culture.

In most organizations, downward influence is thought of as traditional leadership. It can range from bosses who influence by telling others what to do to leaders who work with their direct reports by listening, learning, mentoring, trusting and using objective data to assess performance while ensuring accountability. Effective leadership can create new opportunities, jobs, economic growth and prosperity, while ineffective leadership can have catastrophic consequences. Perhaps that is why a whole industry has grown around teaching leadership, in trying to replicate what has worked rather than understand why it has worked. Barbara

Kellerman, founding director of the Kennedy School's Center for Public Leadership, has challenged the status quo in her book *The End of Leadership*, asserting that the focus should shift from the concept of single heroic leaders to a broader understanding of leadership, followership and context in helping organizations achieve peak performance. Looking at influence within organizations from a broader level may help leaders contextualize their role in achieving desired outcomes.

Peer influence, or what is more commonly known as peer pressure, is often thought of as a negative aspect of interpersonal interactions. In certain contexts that may be true but by reframing the perception of peer influence by asking how it may help organizations it is possible to recognize the potential benefits of positive peer-to-peer influence.

At Kennedy Space Center during the shuttle era, there was a tremendous amount of individual and organizational pride in caring for a fleet of arguably the most complex spacecraft in history. In the orbiter processing facility, team members quickly learned through training and peer feedback that tools had to be returned to the tool board after use to prevent inadvertently leaving a tool inside the spacecraft. People were not afraid to speak up if a tool was missing or if the protocols were not being followed, which then became part of the culture. If an individual did not follow protocols, their colleagues would typically speak up to let them know "This is NASA, that's not the way we do things here." In hospitals where independent audits of hand hygiene often show actual compliance rates of 80 to 90%, peer pressure can play an important role if team members are willing to respectfully speak up when someone is not doing what everyone knows needs to be done.

In organizations where trust is high and communication respectful, peer influence can be a very effective approach to building and maintaining organizational excellence. When there are low levels of trust, poor communication, gossiping or when individuals are blamed, morale is poor, mistakes are not caught, performance is not where it needs to be and there is little, if any, pride in being part of the organization.

Many senior leaders have learned that there are influencers within an organization who do not have designated leadership roles. Their names do not appear on the org. chart, they have no official leadership responsibilities or authority, yet they have significant influence in the organization. These individuals are highly respected by their peers for their competency, their willingness to mentor, their knowledge of corporate culture and their commitment to doing what's right. George Abbey understood the importance of knowing what was really going on. To succeed, leaders need to know what is really happening within the organization. Ideally that information is provided through the regular management channels, but it is easy for some to fall into the trap of telling leaders what is working and not disclose what is not working. Those who worked directly with Mr. Abbey knew he wanted to know what was working, what wasn't working and if that were the case, what the plan was to address the issue with metrics to assess progress.

Jim Wetherbee and I worked together closely when I was the director of the Space and Life Sciences Directorate at Johnson Space Center. In his role as director of flight crew operations, Jim and I both reported directly to George Abbey. We chatted about the process of decision making. Jim recalled:

> It isn't just making the right decisions; you have to make sure that the workforce is giving you the right information to make the right decision. That's another thing that George was an expert at. He didn't just rely on his direct reports to give him the information. He went out and verified it from lower-level people in the organization. So, he knew what was going on and he could encourage people to tell him what they

needed, what he needed to make the decision. When I was working for him, I coined the phrase 'sunshine reports.' Too many bosses only want the sunshine report. George never wanted that, you know, in fact, he would be kind of bored when I was . . . explaining to him what we were doing last week and what we're going to do next week. . . . When I told him things were going well, he gave me the body language indications that he was somewhat uninterested. But when I told him there was an issue or a problem, he perked right up, and he gave me all the positive body language indications that he was really interested in what I was saying. He was essentially rewarding me for searching for vulnerabilities, which is very important in a technical organization if you don't want to have accidents. And even if I didn't have the answer — you know, you've heard the saying don't ever give your boss a problem for which you don't have an answer. Well, I disagree with that. If I didn't understand how to solve a problem, I still told George, here's the issue we're having. And he was very happy that I at least identified the issue and then at least I was going to work on it next week, even if I didn't have a solution. Maybe he had a solution. Maybe he didn't. But at least I identified the vulnerability and we were going to work at it. So that's what he rewarded me for, searching for vulnerabilities.

I had the same experience in my weekly meetings with George and I was never afraid to bring an issue to his attention. Jim recalled, "George Abbey set expectations. He created commitment in the people working below him for human spaceflight, he created a process of accountability."

Leaders are well served to consider the interwoven nature of accountability, power, authority, responsibility and leadership. Many would say it's pretty simple: the person at the top is in charge, they can fire you if they're not happy with you. Your livelihood and that of your family is, to a large degree, in their hands. It's natural to want to please the boss, and it is easy to fall into the trap of doing what people think the boss wants done rather than focusing on what needs to be done. Successful NASA leaders focus on getting teams to do what needs to be done, to do it well and do it safely. It is an environment where outcomes are critical; mission success, operational success is paramount. The NASA leadership moments are defined by how the leader and team worked together to ensure success. That relationship is dependent upon a clear understanding of how leaders and teams work together, yet in some organizations little consideration is given to the subtle differences between power, authority, leadership and accountability.

The typically pyramidal corporate org. chart graphically portrays the distribution of authority. The leader at the top has been given the authority to run the organization. Whether appointed or elected, they have the authority and the responsibility to run the organization. With that authority comes power, the power to do what they believe needs to be done. Some choose to wield that power by using their positional authority and a directive leadership style. Others choose to lead through influence, clearly communicating responsibilities and expectations while mentoring, role modeling and building the capacity of the team. Most, but not all, leaders understand that excessive use of positional authority is not productive.

However, there are situations that require the use of power associated with positional authority to achieve results. Both Dan Goldin and Mike Griffin used their positional authority to replace a number of senior leaders, to change reporting structures and to implement new ways of working in response to budgetary overruns that threatened to cancel the space station program and to the loss of the space shuttle Columbia. Both circumstances required decisive action and, given the changes required, the use of positional authority was appropriate. Doing what is right for the organization is often not popular and fortunately both leaders were able to weather the storm.

Gene Kranz and Glynn Lunney acted decisively in responding to Apollo 13 and succeeded by using influence to leverage the competency of the mission control team and astronauts. George Low characteristically led through influence to achieve the seemingly impossible recovery from the Apollo 1 fire to landing on the Moon in less than three years. On rare occasions when he had no recourse, he would use his positional authority with contractors to achieve the results he needed. There is no magic formula that leaders can use to understand when to use their power and when to use their influence; for most it is learned through mentoring or through experience. That's why it's called the school of hard knocks.

It can be helpful to think of a continuum from authority at one end to influence at the other. The different leadership styles fall on different parts of the continuum with directive and pace-setting styles more aligned with leading through power and authority while visionary, affiliative and coaching/mentoring styles are more aligned with leading through personal influence. It is virtually impossible for a successful leader to use only one leadership style: the goal for leaders is to develop a repertoire of leadership styles that they can select from dependent upon the situation. Strategic planning aligns with visionary leadership while selecting for and building competency requires a coaching/mentoring leadership style — the best leaders understand situationally specific leadership and are flexible in their approach. While many leaders focus on leading through personal influence, a willingness to selectively use power to deal with poorly performing individuals or to do what an organization needs done is important. Fortunately, NASA had many highly effective leaders able to mentor others on that subtle art.

Whatever style the overall leader chooses, everyone in an organization should clearly understand their role and associated responsibilities. Those in leadership roles within the organization should know the scope of their authority, their responsibilities and those of their direct and indirect reports and maintain accountability through regular communication to determine if responsibilities are being met. Regular, respectful, candid communication is the key to success. Experienced leaders want to hear what they need to hear, not what those reporting to them think they want to hear.

It is also not a surprise that in some organizations, individuals do not clearly understand their responsibilities and are not held accountable. Poor communication undermines accountability. Without accountability, quality of work decreases, and no one is sure who has the responsibility to solve the problem. Leaders who "shoot the messenger" will quickly find that they are not hearing about problems until they become so large that they are evident to all. By that point the problems are much harder to solve, which creates even more frustration and anger in the person in charge.

Looking back at close to 60 years of human space exploration, it is clear that there have been times of tremendous achievements interspersed with more difficult times. Much has been written about the organizational culture and decision making that contributed to the losses of Columbia, Challenger and Apollo 1 but there is another aspect to working in a high-stakes operational world that is worth exploring. That is leading at the edge, leading at the boundaries of individual and team comfort zones, developing techniques to control risk while pushing the edge of what's possible.

History would suggest that NASA during the Mercury, Gemini and Apollo programs was an organization that defined, redefined and continued to push the limits of human spaceflight. Within a 10-year period, the agency went from not knowing how to fly in space to landing and successfully returning humans from the Moon. Achieving those results took courage, commitment, competency and a willingness to work at the limits of individual and organizational comfort zones to achieve what many observers thought was impossible. After success, the obvious question is what's next?

The answer is not always straightforward; government programs are impacted by national policies, funding and competing political priorities. The Vietnam War, societal unrest and budgetary constraints all contributed to President Nixon's decision not to build on the success of Apollo by implementing bold, expensive space programs. Arguably public support alone for continued spending on space exploration would have been a challenge. But for those NASA engineers and astronauts who had spent a decade pushing themselves to be the best they could be, it would not be surprising for them to feel like a gold medal–winning Olympian after the Olympics. What's next? What will be as meaningful as what they have just done?

Successfully testing personal limits, pushing to set new records, is tough but rewarding. It is likely that anything less than sending humans to Mars would have been considered somewhat mundane after the achievements of the Apollo era. While it would be a stretch to say NASA became complacent in the Skylab and Apollo-Soyuz programs, the missions of the seventies were well within the comfort zone of the team. Naturally there were retirements as the program shifted focus from high-reward, high-risk human missions beyond Earth orbit to building a space shuttle and ultimately a space station. Did those changes affect the sustained impact of the lessons learned on how teams work safely within and at the limits of their knowledge and capacity?

Some might argue that over time NASA shifted from the operational priorities associated with mission success to more bureaucratic priorities of administrative success overseeing the contracts, costs and deliverables associated with building the space shuttle. The first shuttle flight, STS-1 on April 12, 1981, was a remarkable demonstration of a complex new multipurpose spacecraft that would increase accessibility and reduce the costs of getting to low Earth orbit. Over the next five years the rate of shuttle flights increased and, while there were operational issues, the overriding schedule pressure to launch commercial payloads and meet other programmatic needs shifted the agency priorities. While some recognized that human spaceflight is inherently risky and the shuttle program was still an experimental test program exploring the new unknowns, others felt that the shuttle was capable of being flown as a commercial space-craft well within the comfort zone of the agency.

In hindsight, it is perhaps easy to suggest that there was scope creep away from data-driven decisions characteristic of exploring test limits towards experience-driven decisions derived from doing the same thing frequently. The 21st century has brought a unique set of unforeseen circumstances that affect global and national economies. Companies are focused on organizational agility, piv-oting to meet new consumer needs or new ways of doing business. That has taken many organizations to the limits of their comfort zones as they try to develop new innovative technologies or new ways of doing business to sustain and grow. The NASA experiences of working at the edge of personal and corporate comfort zones are as relevant today as they were in the past. Flourishing in uncer-tain times is the result of hiring the best, continued organizational learning and innovative thinking, building high-trust organizations, listening to expert opinions while making data-driven decisions and thoughtful leadership.

Like any organization, NASA has had its challenges and suc-cesses, but few would say they were not proud to have worked there. Sitting in the recently restored Apollo mission control, it is impossible not to feel the pride of mission success, the tension of the many tough moments that occurred during the missions — and imagine what it would have been like to have been part of one of those teams. Where the agency achieved its finest moments, leaders relied on the competency of the teams they had built, the individuals they had helped become the experts who had mastered the complexities of sending humans to space. They trusted the teams implicitly but also used probing questions and data, as well as independent sources of information to verify that they were ready to take on the most daunting of challenges, sending humans to space.

APPENDIX

Synthesis Group Report[1]

Guidelines

1. Establish crew safety as the number one priority.
2. Have clean lines of management authority and responsibility for all elements of the program. Ensure that one organization or prime contractor is clearly in charge.
3. Establish realistic program milestones that provide clear entry and exit criteria for the decision process and create useful capabilities at each step.
4. Ensure that the administration and Congress clearly understand the technical and programmatic risks and realistic costs of the space exploration initiative.
5. Mandate simple interfaces between subsystems and modules.
6. Make maximum use of modularity over the life of the program to maintain flexibility. Successive missions should build on the

---

1. Appendix A-8 legacies, Synthesis Group report, 1991, 130.

capabilities established by prior ones. Provide the capability to incorporate new technology as required.

7. Press the state-of-the-art and technology when required and/or when technological opportunities are promising with acceptable risk.

8. Ensure optimum use of men in the loop. Don't burden man if a machine can do it as well or better and vice versa.

9. Limit development times to no more than 10 years; if it takes longer, the cost goes up and commitment goes down.

10. Focus technology development toward programmatic needs.

11. Minimize or eliminate on-orbit assembly requiring extra-vehicular activity.

12. Minimize mass to lower Earth orbit to reduce cost.

13. Have redundant primary and separate backup systems. Design in redundancy versus heavy reliance on onboard/onsite maintenance.

14. Hire good people, then trust them.

Pitfalls

1. Establishing requirements that you will be sorry for. For instance, wish lists being treated as requirements and allowing requirements to creep.

2. Trying to achieve a constituency by promising too much to too many and lowballing the technical and financial risks.

3. Committing to interminable studies and technology demonstrations without a firm commitment to execute a real program.

4. Not establishing configuration controls/baselines as soon as possible; e.g., weight and electrical power requirements.

5. Allowing software to run unchecked and become a program constraint rather than a supporting element.

6. Setting up agreements for development of program elements that are not under direct program management control.

7. Not saying we were wrong when we were wrong.

ACKNOWLEDGMENTS

I would like to thank my co-author, Elizabeth Howell, for her hard work, patience, persistence and support. Thanks for fielding my many calls and your passion and optimism. I really appreciated your quick responses with the writing and many edits! It has been a pleasure working with you and I am looking forward to our continued collaboration.

It is with profound humility that I would like to thank all of the current and former NASA executives who shared time from their busy schedules with me. My leadership journey has been enriched immensely from our conversations and I am deeply grateful for your insights. It was a challenge writing about NASA's leadership history. For those who enjoy this book, we hope it is of some assistance to your leadership endeavors. Some may wonder why various leaders and stories were not included. There are many incredible leadership stories and moments from NASA's history. These are the stories that resonated with us and we hope that you enjoyed them, and that they have helped illustrate the many facets of leadership, followership and teamwork for you.

To Mac Evans, Dan Goldin, Mike Griffin, Joe Rothenberg, George Abbey, Bill Gerstenmaier, Beak, Bill Parsons, Dave King, Randy Brinkley, Janet Kavandi, Jim Wetherbee and Kathy Nado, it was an honor to have worked with you. We shared a number of late nights, long hours, challenging moments, times to celebrate and a time to mourn. Those memories will stay with me forever. I would also like to thank Dr. Arnauld Nicogossian and Dr. Rich Williams, both former NASA chief medical officers, for being excellent mentors, colleagues and great friends. To the space and life sciences team, thanks for your outstanding work and commitment to excellence, crew health and safety. To John and Diana, I couldn't have done it without you!

To George Abbey, thank you for your wisdom, guidance, humility and your friendship. I am deeply appreciative of the faith you had in my leadership skills and the opportunity you gave me. Your commitment to always doing the right thing is an important lesson for everyone, thanks for your mentorship and sharing it with me.

To my colleagues at the Canadian Space Agency, it was an honor to work with you and be able to represent Canada in space on my two spaceflights. Special thanks to Mac Evans for your leadership that made it all possible. Your legacy has ensured that Canada is one of the world's outstanding spacefaring nations. To Jean-Marc and the OSM team thanks for keeping me healthy and caring for my family when I was in space and in the ocean.

Thanks to my colleagues in the astronaut office for your leadership, followership and excellent teamwork. To the crews of STS-90 and 118, NEEMO-7 and 9, through teamwork we triumphed.

Special thanks to our publisher Jack David and the ECW team. We really appreciate your excitement about this project, your unwavering support and the many hours of reviewing manuscripts.

To my wife, Cathy, and Olivia, Evan and Theo, thanks for your support, patience and understanding.

To my good friend Scott Haldane, former President and CEO of YMCA Canada, thanks for your comments on the manuscript.

Finally, thanks to our readers for your interest in leadership and teamwork. The journey is a lifetime endeavor with many ups and downs, twists and turns and is filled with continuous opportunities to learn. From first-hand experience I know it is not for the faint of heart but the accomplishments along the way make up for the challenges. Congratulations on what you have achieved and what you will achieve.

DAVID R. WILLIAMS OC OOnt MSc MD CM
FCFP FRCPC FRCP DSC (HON.) LLD (HON.)

I (Elizabeth Howell) would like to first of all thank my co-author, Dave Williams, for all of his insights about leadership that helped inform the writing of this book. He is such a humble person who probably will be reluctant to let me say this about him, but in any case: Dave has an admirable hard-working attitude and enthusiasm, even amid the intense deadlines and work that a lengthy research project like this entails. Dave has a unique combination of skills from being a CEO, NASA executive, Canadian astronaut, leader of a NASA aquanaut mission and a medical doctor, and thus brings experience and connections from many different fields into this book. I couldn't have asked for a better co-writing partner and am continually humbled by the effort he brought to this writing partnership. Thanks, Dave, for letting me join this special mission.

I am also grateful to all of the space leaders and leaders in other fields who shared their experience with us in interviews and in reading early drafts of this book, as they provided so much help in shaping the focus of our tale. We gained a lot of insights from dozens of hours of interviews that they generously gave to us. We also leaned on the experience of many space leaders who provided interviews in books and oral interviews, and we thank all the hardworking folks behind the scenes who brought those tales into the public sphere for us to draw from. Dave and I hope that we have done their stories justice, as we greatly admire each and every one of the people who generously gave their time to us.

Thank you also to ECW Press for their ongoing support of my writing about space, and for providing great comments on earlier drafts that made the difference in presenting this story for a leadership audience. ECW willingly shared time in numerous e-mails and phone calls to discuss the vision of this book, and their team was very enthusiastic in all aspects of the book from conception to publication to publicity. Thanks for understanding the approach Dave and I wanted to take.

Special thanks to my husband J for understanding (once again) that special drive it takes to write a book, and for providing advice on how to approach the complex topic of leadership, drawing from his own award-winning work experience. He has seen me through four complex books through our years together. J not only supports these efforts but encourages me to continue because he believes in me — I am ever so grateful for that.

Finally, thank you to all those who will read this book; we hope that the NASA-focused leadership tales here will help you in your own leadership endeavours.

ELIZABETH HOWELL B. JOUR MSC PHD

The authors also wish to pay tribute to Glynn Lunney, a vital NASA leader who was hired when the agency was formed in 1958 and played key roles in the Gemini, Apollo and Apollo-Soyuz programs. Lunney was an outstanding leader who exemplified humility in his many leadership roles. He passed away in March 2021 during the final editing of this book and will always be remembered as one of the great NASA leaders. The authors express condolences to his family, thanking them for his service to the space community.

NOTES

Chapter 1

1. Charles Fishman, "The birth of the electronic beep, the most ubiquitous sound design in the world," Fast Company, last modified June 8, 2019, https://www.fastcompany .com/90361076/the-birth-of-the-electronic-beep-the-most-ubiquitous-sound-design-in-the-world.
2. Peter Pindjak, "The Eisenhower Administration's Road to Space Militarization" (Master's diss., Evanston, IL: Northwestern State University, 2009), 48.
3. James L. Schefter, *The Race: The Uncensored Story of How America Beat Russia to the Moon* (New York: Doubleday, 1999).
4. NASA History Office, "60 Years Ago: Vanguard Fails to Reach Orbit," last modified January 20, 2018, https://www .nasa.gov/feature/60-years-ago-vanguard-fails-to-reach-orbit.
5. Keith Glennan, interview by Martin Collins and Dr. Allan Needell, NASA Johnson Space Center Oral History Project (JSC OHP), episode 5 (May 29, 1987), audio, 4:19.

6. Glennan, interview, 4:19.
7. Glennan, interview, 4:19.
8. Glennan, interview, 6:19.
9. Roger D. Launius, "Leaders, Visionaries and Designers," NASA History Report. https://www.nasa.gov/50th/50th_magazine/leaders.html.
10. Glennan, interview, 8:19.
11. Glennan, interview, 14:19.
12. Glennan, interview, 15:19.
13. Owen E. Maynard, interview by Carol Butler, JSC OHP, April 21, 1999, https://historycollection.jsc.nasa.gov/JSCHistoryPortal/history/oral_histories/MaynardOE/MaynardOE_4-21-99.htm.
14. Chris Gainor, *Arrows to the Moon*, (Burlington, ON: Apogee Books, 2001), 37.
15. Launius, "Leaders, Visionaries and Designers."

Chapter 2

1. Stever Report quoted in Richard Jurek's *The Ultimate Engineer: The Remarkable Life of NASA's Visionary Leader George M. Low* (Lincoln, NE: University of Nebraska Press, 2019, e-book), 67.
2. Robert R. Gilruth quoted in Glen E. Swanson's *Before This Decade is Out, Personal Reflections of the Apollo Program*, NASA SP-4223, 1999, 66.
3. Robert R. Gilruth quoted in Richard Jurek's *The Ultimate Engineer* (e-book), 44.
4. Jurek, *The Ultimate Engineer* (e-book), 50.
5. George M. Low quoted in Richard Jurek's *The Ultimate Engineer* (e-book), 48.
6. Manfred "Dutch" von Ehrenfried, "Appendix 1" in *The Birth of NASA* (Springer Praxis Books, 2016), 179.

7. National Academy of Sciences, Chris Kraft quoted in the *Biographical Memoirs: Volume 84* (Washington, DC: The National Academies Press, 2004), 93.

8. Christopher C. Kraft, Jr., *Flight: My Life in Mission Control* (New York: Penguin Group, 2001), 67.

9. Kraft, *Flight*, 67.

10. McKinsey & Company, definition of "organizational agility."

11. Kraft, *Flight*, 67.

12. Kraft, *Flight*, 71.

13. National Weather Service, "Summary of the January 18 — 20th 1961 Nor'easter," https://www.weather.gov/rlx/jan61.

14. Robert R. Gilruth, interview by David DeVorkin and John Mauer, episode 6, JSC OHP, March 2, 1987, https://airandspace.si.edu/research/projects/oral-histories/TRANSCPT/GILRUTH6.HTM.

15. Gilruth, interview, 1987.

Chapter 3

1. Gilruth, "From Wallops Island to Project Mercury, 1945 - 1958: A memoir." *Essays on the History of Rocketry and Astronautics*, Vol. 2. NASA Technical Research Server, 445, https://ntrs.nasa.gov/citations/19770026126.

2. Gilruth, "memoir," 471.

3. Eugene F. Kranz, interview by Roy Neal, JSC OHP, March 19, 1998, https://historycollection.jsc.nasa.gov/JSCHistoryPortal/history/oral_histories/KranzEF/KranzEF_3-19-98.htm.

4. Kranz, interview, 1998.

5. Kranz, interview, 1998.

6. Kranz, interview, 1998.

7. James M. Grimwood, Loyd S. Jr. Swenson, and Charles C. Alexander, *This New Ocean: A History of Project Mercury*,

NASA SP-4201 in The NASA History Series (Washington, DC: NASA History Office, 1998).

8. Kraft or Kranz.

9. Jeffrey Pfeffer and Robert L. Sutton, *The Knowing-Doing Gap: How Smart Companies Turn Knowledge into Action* (Brighton, MA: Harvard Business Review Press, 1999), 29.

10. Pfeffer and Sutton, *The Knowing-Doing Gap*, 42.

11. John Glenn with Nick Taylor, *John Glenn: A Memoir* (New York: Bantam Books, 1999), 260.

12. Glenn, *Memoir*, 269.

13. Christopher C. Kraft, Jr., *Flight: My Life in Mission Control* (New York: Penguin Group, 2001), 158.

14. Kraft, *Flight*, 158.

15. Kraft, *Flight*, 160.

16. Kranz, 1998.

17. David J. Shayler, *Disasters and Accidents in Manned Spaceflight* (Springer Praxis, 2000), 353.

18. Shayler, 355.

19. Kraft, interview by Rebecca Wright, JSC OHP, August 6, 2012, https://historycollection.jsc.nasa.gov/JSCHistoryPortal/history/oral_histories/KraftCC/KraftCC_8-6-12.htm.

20. Kranz, interview, 1998.

21. Kranz, interview, 1998.

22. Kranz, interview, 1998.

23. Kranz, interview, 1998.

24. Kranz, interview, 1998.

Chapter 4

1. David R. Williams, "The Apollo 1 Tragedy," NASA, 2018, https://nssdc.gsfc.nasa.gov/planetary/lunar/apollo1info.html.

2. Betty Grissom and Henry Still, *Starfall* (New York: Thomas Y. Crowley Company, 1974), 24.

3. Apollo 204 Review Board, "Findings, Determinations and Recommendations," NASA Historical Reference Collection, NASA History Office, NASA Headquarters. https://history .nasa.gov/Apollo204/find.html.

4. Frank Borman, "Hearings Before the Subcommittee on NASA Oversight of the Committee on Science and Astronautics" (U.S. House of Representatives, Ninetieth Congress, First Session: April 10, 11, 12, 17, 21; May 10, 1967), 85.

5. Borman, 85.

6. Apollo 204 Review Board, "Description of Test Sequence and Objectives: Events from Initiation of the Plugs-Out Test Until the T-10 Minute Hold," NASA Historical Reference Collection NASA History Office, NASA Headquarters, https://history.nasa.gov/Apollo204/desc.html.

7. Nassim Nicholas Taleb, *The Black Swan: The Impact of the Highly Improbable* (New York: Random House, 2007), xx.

8. Frank Borman, interview by Catherine Harwood, JSC OHP, April 13, 1999, https://historycollection.jsc.nasa.gov/ JSCHistoryPortal/history/oral_histories/BormanF/ Bormanff_4-13-99.htm.

Chapter 5

1. Courtney G. Brooks, James M. Grimwood, and Loyd S. Jr. Swenson, *Chariots for Apollo: The NASA History of Manned Lunar Spacecraft to 1969* (North Chelmsford, MA: Courier Corporation, 2012), 11.

2. Glen E. Swanson (ed.), "George M. Low (1926-1984)" in *Before This Decade Is Out*, NASA SP-4223, NASA History Series, 2012. https://history.nasa.gov/SP-4223/ch13.htm.

3. Bob Granath, "Gemini XII Crew Masters the Challenges of Spacewalks," NASA, 2016, https://www.nasa.gov/feature/ gemini-xii-crew-masters-the-challenges-of-spacewalks.

4. Richard Jurek, *The Ultimate Engineer: The Remarkable Life of NASA's Visionary Leader George M. Low.* (Lincoln, NE: University of Nebraska Press, 2019), 131.

5. Daniel H. Pink, *Drive: The Surprising Truth About What Motivates Us* (New York: Riverhead Books, 2011).

Chapter 6

1. Eric M. Jones (ed.), "The First Lunar Landing," Apollo 11 Lunar Surface Journal, last modified May 10, 2018, https://www.hq.nasa.gov/alsj/a11/a11.landing.html.

2. Wayne Hale, "Nexus of Evil," Wayne Hale's Blog, February 16, 2010, https://blogs.nasa.gov/waynehalesblog/2010/02/16/post_1266353065166/.

3. Jones, 2018.

4. Francis E. "Frank" Hughes, interview by Rebecca Wright, JSC OHP, September 17, 2013, https://historycollection.jsc.nasa.gov/JSCHistoryPortal/history/oral_histories/HughesFE/HughesFE_9-17-13.htm.

5. Jones, Eric M. (ed.) "Building on Experience," Apollo Lunar Surface Journal, last modified June 10, 2014, https://www.hq.nasa.gov/alsj/apollo.precurs.html.

6. Thomas P. Stafford, interview by William Vantine, JSC OHP, October 15, 1997, https://historycollection.jsc.nasa.gov/JSCHistoryPortal/history/oral_histories/StaffordTP/StaffordTP_10-15-97.htm.

7. John R. Garman, interview by Kevin M. Rusnak, JSC OHP, March 27, 2001, https://historycollection.jsc.nasa.gov/JSCHistoryPortal/history/oral_histories/GarmanJR/GarmanJR_3-27-01.htm.

Chapter 7

1. Richard Hollingham, "The Switch That Saved a Moon Mission from Disaster," BBC, 2019, https://www.bbc.com/future/article/20191108-the-switch-that-saved-a-moon-mission-from-disaster.
2. Hollingham, "The Switch That Saved a Moon Mission."
3. Charles J. Conrad and Alan B. Shepard, "Ocean of Storms and Fra Mauro" in *Apollo Expeditions to the Moon*, NASA SP-350, 1975. https://history.nasa.gov/SP-350/ch-12-1.html.
4. John W. Aaron, Kevin M. Rusnak, JSC OHP, January 18, 2000, https://historycollection.jsc.nasa.gov/JSCHistoryPortal/history/oral_histories/AaronJW/AaronJW_1-18-00.htm.
5. Eric M. Jones (ed.), "TV Troubles" in Apollo 12 Lunar Surface Journal, last modified August 4, 2017. https://www.hq.nasa.gov/alsj/a12/a12.tvtrbls.html.
6. Jones, "TV Troubles."
7. Jones, "TV Troubles."
8. Richard F. Gordon, Jr., interview by Michelle Kelly, JSC OHP, October 17, 1997, https://historycollection.jsc.nasa.gov/JSCHistoryPortal/history/oral_histories/GordonRF/GordonRF_10-17-97.htm.

Chapter 8

1. James A. Lovell, Jr., interview by Ron Stone, JSC OHP, May 25, 1999, https://historycollection.jsc.nasa.gov/JSCHistoryPortal/history/oral_histories/LovellJA/LovellJA_5-25-99.htm.
2. Williams, David R. "The Apollo 13 Accident," NASA, 2016, https://nssdc.gsfc.nasa.gov/planetary/lunar/ap13acc.html.
3. Eugene F. Kranz, interview by Jennifer Ross-Nazzal, JSC OHP, December 7, 2011, https://historycollection.jsc.nasa

.gov/JSCHistoryPortal/history/oral_histories/KranzEF/
KranzEF_12-7-11.htm.

4. Eugene F. Kranz, interview by Rebecca Wright, JSC OHP,
    January 8, 1999, https://historycollection.jsc.nasa.gov/
    JSCHistoryPortal/history/oral_histories/KranzEF/KranzEF_
    1-8-99.htm.

5. Glynn S. Lunney, "Thought Leader Series: An Evening with
    Glynn Lunney," Space Center Houston, November 15, 2018,
    video, https://spacecenter.org/video-thought-leader-series-an-
    evening-with-glynn-lunney/.

6. Ken Mattingly, interview by Andrew Chaikin in *Voices from
    the Moon* (New York: Penguin Group, 2009), 139.

Chapter 9

1. Edward Clinton Ezell and Linda Neuman Ezell, "Prologue"
    in *The Partnership: A History of the Apollo-Soyuz Test
    Project*, NASA SP-4209, NASA History Series, 1978, https://
    www.hq.nasa.gov/office/pao/History/SP-4209/toc.htm.

2. Glynn S. Lunney, interview by Carol Butler, JSC OHP,
    March 30, 1999, https://historycollection.jsc.nasa.gov/
    JSCHistoryPortal/history/oral_histories/LunneyGS/
    LunneyGS_3-30-99.htm.

3. Thomas P. Stafford, interview by William Vantine, JSC OHP,
    October 15, 1997, https://historycollection.jsc.nasa.gov/
    JSCHistoryPortal/history/oral_histories/StaffordTP/
    StaffordTP_10-15-97.htm.

4. Culbertson, Frank L. "What's in a name?" Paper Submitted
    to 10th Congress of The Association of Space Explorers,
    October 3, 1996, https://spaceflight.nasa.gov/history/
    shuttle-mir/references/to-r-documents-mirmeanings.htm.

5. Sagdeev, Roald. "United States-Soviet Space Cooperation

during the Cold War." NASA, 2008. https://www.nasa.gov/
50th/50th_magazine/coldWarCoOp.html.

Chapter 10

1. Steve Garber (ed.), "Transcript of the Challenger Crew
    Comments from the Operational Recorder," NASA, 2003,
    https://history.nasa.gov/transcript.html.
2. Jay H. Greene, interview by Sandra Johnson, JSC OHP,
    December 8, 2004, https://historycollection.jsc.nasa.gov/
    JSCHistoryPortal/history/oral_histories/GreeneJH/
    GreeneJH_12-8-04.htm.
3. Diane Vaughan, *The Challenger Launch Decision: Risky
    Technology, Culture, and Deviance at NASA* (Chicago:
    University of Chicago Press, 1997).
4. The Rogers Commission, "The Contributing Cause of the
    Accident," Report of the Presidential Commission on the
    Space Shuttle Challenger Accident. NASA, 1986, https://
    history.nasa.gov/rogersrep/v1ch5.htm.
5. Gerald W. Smith, interview by Rebecca Wright, JSC OHP,
    May 12, 2011, https://historycollection.jsc.nasa.gov/
    JSCHistoryPortal/history/oral_histories/STS-R/SmithGW/
    SmithGW_5-12-11.htm.

Chapter 11

1. Valerie Neal, "Remembering Challenger 25 Years Later,"
    Smithsonian Institution National Air and Space Museum,
    https://airandspace.si.edu/stories/editorial/remembering-
    challenger-25-years-later.
2. The Rogers Commission, "The Contributing Cause of the
    Accident," Report of the Presidential Commission on the

Space Shuttle Challenger Accident, NASA, 1986, https://history.nasa.gov/rogersrep/v1ch5.htm.

3. David Ignatius, "Did The Media Goad NASA Into the Challenger Disaster?" *The Washington Post*, 1986, https://www.washingtonpost.com/archive/opinions/1986/03/30/did-the-media-goad-nasa-into-the-challenger-disaster/e0c8669d-a809-4c8d-a4f8-50652b892274/.

4. Arnold D. Aldrich, "Challenger," NASA Program Report, 2008, 3.

5. Robert L. Crippen, interview by Rebecca Wright, JSC OHP, May 26, 2006, https://historycollection.jsc.nasa.gov/JSCHistoryPortal/history/oral_histories/CrippenRL/CrippenRL_5-26-06.htm.

6. Aldrich, interview by Rebecca Wright, JSC OHP, April 28, 2008, https://historycollection.jsc.nasa.gov/JSCHistoryPortal/history/oral_histories/AldrichAD/AldrichAD_4-28-08.htm.

7. Aldrich, "Challenger," internal NASA report, JSC OHP, August 27, 2008.

Chapter 12

1. George M. Low quoted in Richard Jurek's *The Ultimate Engineer: The Remarkable Life of NASA's Visionary Leader George M. Low* (Lincoln, NE: University of Nebraska Press, 2019, e-book), 218.

2. Jurek, *The Ultimate Engineer*, 160.

3. Augustine Committee, "Report of the Advisory Committee On the Future of the U.S. Space Program, 1990," https://history.nasa.gov/augustine/racfup1.htm.

4. Kathy Sawyer, "Truly Fired as NASA Chief," *The Washington Post*, February 13, 1992.

5. Richard Truly, in Michael Cassut's *The Astronaut Maker* Chicago Review Press, 2018.

6. Ken Mattingley in Michael Cassut's *The Astronaut Maker*, 92.

7. George Abbey, interview by David R. Williams (DRW), 2020.

8. Jurek, *The Ultimate Engineer*, 65.

9. Abbey, DRW interview.

10. John Kanengieter and Aparna Rajagopal-Durbin, "Wilderness Leadership — on the Job," *Harvard Business Review*, April 2012.

11. Abbey, DRW interview.

12. Cassut, *The Astronaut Maker*, 330-331.

13. Abbey, DRW interview.

14. Cassut, *The Astronaut Maker*, 334.

15. Andrew Chaikin, "George Abbey: NASA's Most Controversial Figure," Space.com, 2001.

16. Mark Albrecht, interview by Rebecca Wright, JSC OHP, April 20, 1999, https://historycollection.jsc.nasa.gov/JSCHistoryPortal/history/oral_histories/Shuttle-Mir/AlbrechtMJ/AlbrechtMJ_4-20-99.htm.

Chapter 13

1. Ben Evans, "Techno-turkey: Remembering Hubble's Vision Troubles, 30 Years On," Astronomy.com, June 26, 2020, https://astronomy.com/news/2020/06/techno-turkey-remembering-hubbles-vision-troubles-30-years-on.

2. *Orlando Sentinel*, "Lost in Space" May 17, 1990, https://www.orlandosentinel.com/news/os-xpm-1990-05-17-9005170447-story.html.

3. Daniel S. Goldin, interview by DRW, 2020.

4. Goldin, DRW interview.

5. W. Henry Lambright, "Transforming Government: Dan Goldin & the Remaking of NASA," PWC leadership study, 2001.

6. Goldin, DRW interview.

7. Jim Wetherbee, interview by DRW, 2020.

8. Goldin, DRW interview.

9. Goldin, DRW interview.

10. Gregg Easterbrook, "Don't Replace the Chief Shaking up Space Agency: NASA's Daniel Goldin is doing what has been long overdue-making the agency cost-effective and flexible— and he should be kept on the job." *Los Angeles Times*, November 29, 1992.

11. Easterbrook, "Goldin," 1992.

12. Sean Holton, "NASA Countdown: Goldin to Remain?" *Orlando Sentinel*, December 10, 1992.

13. Glen E. Swanson, "Worms and Wings, Meatballs and Swooshes: NASA Insignias in Popular Culture," The Space Review, 2020. https://www.thespacereview.com/article/3947/1.

14. Swanson, "NASA insignias."

15. Kathy Sawyer, "White House to Retain Goldin as NASA Chief," *The Washington Post*, June 24, 1993.

16. Marcia Dunn, "NASA's Goldin Boy Wants to Boldly Go Where No Man Has Gone Before: Space: He's outspoken, determined and in charge. Dan Goldin is shooting for the stars with plans for a space station and a mission to Mars," *Los Angeles Times*, July 17, 1994.

17. Dunn, "NASA's Goldin Boy."

Chapter 14

1. Daniel S. Goldin, interview by DRW, 2020.

2. Gregg Easterbrook, "Don't Replace the Chief Shaking up Space Agency: NASA's Daniel Goldin is doing what has been long overdue-making the agency cost-effective and flexible — and he should be kept on the job," *Los Angeles Times*, November 29, 1992.

3. Cassut, *The Astronaut Maker*, 342.

4. Sawyer, "Retain Goldin As NASA Chief."

5. Cassut, 352.

6. Cassut, 353.

7. Michael Mott, interview by Rebecca Wright , JSC OHP, April 23, 1999, https://historycollection.jsc.nasa.gov/JSCHistoryPortal/history/oral_histories/Shuttle-Mir/MottM/MottM_4-23-99.htm.

8. Cassut, 357.

9. Cassut, 358.

10. Thomas P. Stafford, "Appendix: Legacies" in *America at the Threshold*, Synthesis Group report, May 6, 1961, https://history.nasa.gov/staffordrep/main_toc.PDF.

11. Stafford, "Appendix: Legacies," *America at the Threshold*, Synthetic Group.

12. Stafford, "Appendix: Legacies," *America at the Threshold*, Synthetic Group.

13. Andrew Chaikin, "George Abbey: NASA's Most Controversial Figure," Space.com, February 26, 2001.

Chapter 15

1. William (Mac) Evans, interview by DRW, 2020.

2. Roald Sagdeev and Susan Eisenhower, "United States–Soviet Space Cooperation during the Cold War," 2011, https://www.nasa.gov/50th/50th_magazine/coldWarCoOp.html.

3. Evans, DRW interview.

4. Evans, DRW interview.

5. Sagdeev and Eisenhower, "States–Soviet Space Cooperation."

6. UPI Article, "Challenger Disaster: World Reacts in Sorrow," U.S. News, January 28, 2016.

7. UPI Article, "Challenger Disaster."

8. Goldin, interview.

9. Sarah Y. Keightly, "Chuck Vest to Head Committee," *MIT Technical Review* 113(16):1, 1993.

10. The American Presidency Project, "Statement on the Space Station Program by President Bill Clinton," June 17, 1993.

11. Evans, DRW interview.

12. Albrecht, interview. (See Ch. 12, note 16.)

13. American Presidency Project, 1993.

14. Goldin, DRW interview.

15. Albrecht, interview.

16. Goldin, DRW interview.

17. William (Bill) Gerstenmaier, interview by DRW, 2020.

18. Kathy Sawyer, "Docking Crash Cripples Mir Space Station," *The Washington Post*, June 26, 1997.

19. Clay Morgan, "NASA History of Shuttle MIR, The United States and Russia Share History's Highest Stage," The NASA History Series, NASA SP-2001-4225, 116.

20. Morgan, 116.

21. Morgan, 116.

22. Keith Cowing, "Dan Goldin Did It His Way," SpaceRef, December 5, 2001.

Chapter 16

1. Northrop Grumman v. United States of America, Contract Formation; Consideration 10 U.S.C. No. 97-359C, April 7, 2000.

2. James F. Peltz and Greg Miller, "Boeing Wins Bid on Space Station: Technology: NASA also singles out Houston as command center for streamlined project. Contract may hurt Southland's McDonnell Douglas and Rockwell," *Los Angeles Times*, August 18, 1993.

3. Carolyn Huntoon, interview by Rebecca Wright, JSC OHP, June 5, 2002, https://historycollection.jsc.nasa.gov/

JSCHistoryPortal/history/oral_histories/HuntoonCL/
HuntoonCL_6-5-02.htm.

4. Huntoon, interview.

5. Cassut, *The Astronaut Maker*, 368.

6. Mott, interview. (See Ch. 14, note 7.)

7. Huntoon, interview.

8. Jim Wetherbee, interview by DRW, 2020.

9. Wetherbee, DRW interview.

10. Wetherbee, DRW interview.

11. Huntoon, interview.

12. Huntoon, interview.

13. Goldin, DRW interview.

14. Randy Brinkley, interview by Rebecca Wright, JSC OHP, February 23, 2016, https://historycollection.jsc.nasa.gov/ JSCHistoryPortal/history/oral_histories/ISS/BrinklelyRH/ BrinkleyRH_2-23-16.htm.

15. Brinkley, interview.

16. Abbey, DRW interview.

17. Abbey, DRW interview.

18. Abbey, DRW interview.

Chapter 17

1. Jerry Linenger, "Space: Astronaut Jerry Linenger Is Interviewed after Fire in Space," interview by The Associated Press, AP Archive, July 21, 2015, video, https://www. youtube.com/watch?v=gKbHhzPJOVM.

2. Linenger, Mir-23 Mission Interviews, NASA, week of March 21, 1997.

3. Linenger, Mir interview.

4. "Preparing for Long Duration Spaceflight," NASA Expeditionary Training Syllabus, 7.

5. Linenger, Mir interview.
6. Michael Foale, interview by Rebecca Wright, JSC OHP, June 16, 1998, https://historycollection.jsc.nasa.gov/ JSCHistoryPortal/history/oral_histories/Shuttle-Mir/ FoaleCM/FoaleCM_6-16-98.htm.
7. Foale, interview.
8. Kathy Sawyer, "Astronaut Foale Recounts Collision at Space Station," *The Washington Post*, October 30, 1997.
9. Foale, interview.
10. Foale, interview.
11. Dorothy Winsor, "Communication Failures Contributing to the Challenger Accident: An Example of Technical Communicators," *IEEE Transactions on Professional Communication* 31, no. 3 (1998): 101-107.
12. Alan J. MacDonald, "Space Shuttle Challenger Disaster," interview by the American Society of Civil Engineers, November 18, 2015, video, https://www.youtube.com/ watch?v=QbtY_WI-hYI.
13. Appendix A-8 legacies, Synthesis Group report, 1991, 130. (See Appendix A.)

Chapter 18

1. Jim Wetherbee, *Controlling Risk in a Dangerous World* (New York: Morgan James Publishing, 2016).
2. Wetherbee, DRW interview.
3. Cassut, *The Astronaut Maker*, 190.

Chapter 19

1. Edward Tufte, "Beautiful Evidence," presentation at Intelligence Squared, London, UK, May 19, 2010.
2. Report of Columbia Accident Investigation Board (CAIB),

August 26, 2003, 162, htttps://www.nasa.gov/Columbia/home/CAIB_Vol1.html.

3. Geoff Brumfeil, "Total Failure: When the Space Shuttle Didn't Come Home," *All Things Considered*, NPR, May 17, 2017, audio, https://www.npr.org/2017/05/17/527052122/total-failure-when-the-space-shuttle-didnt-come-home.

4. Brumfeil, NPR interview.

5. CAIB Report, 2003, 9.

6. CAIB Report, 2003, 9.

7. Diane Vaughan, *The Challenger Launch Decision: Risky Technology, Culture, and Deviance at NASA* (Chicago, IL: University of Chicago Press, 1996).

8. CAIB Report, 191.

9. CAIB Report, 193.

10. Mike and Becky Griffin, interview by DRW, 2020.

11. Griffin, DRW interview.

12. Griffin, DRW interview.

13. Griffin, DRW interview.

14. George W. Bush, "Remarks by the President at the Memorial Service in Honor of the STS-107 Crew," NASA Lyndon B. Johnson Space Center, February 4, 2003.

Chapter 20

1. Joe Rothenberg, interview by DRW, 2020.

2. Rothenberg, DRW interview.

3. Rothenberg, DRW interview.

4. Randy Brinkley, interview by Rich Dinkel, JSC OHP, January 25, 1998, https://historycollection.jsc.nasa.gov/JSCHistoryPortal/history/oral_histories/BrinkleyR/BrinkleyR_1-25-98.htm.

5. Rothenberg, DRW interview.

6. Denise Chow, "Saving Hubble: Astronauts Recall 1st Space Telescope Repair Mission 20 Years Ago," Space.com,

December 2, 2013, https://www.space.com/23640-hubble-space-telescope-repair-anniversary.html.

Chapter 21

1. William (Bill) Gerstenmaier, interview by DRW, 2020.
2. Gerstenmaier, DRW interview.
3. Gerstenmaier, DRW interview.
4. Gerstenmaier, "William Gerstenmaier on Lessons Learned from Large NASA Projects," interview by NASA APPEL Knowledge Services, December 6, 2018, video, https://www.youtube.com/watch?v=ymPnLUT196Q.
5. Meghan Bartels, "NASA Administrator Promises Investigation into Astronauts' Emergency Landing After Soyuz Failure," October 11, 2018, https://www.space.com/42098-soyuz-rocket-launch-abort-nasa-chief-statement.html.
6. Gerstenmaier, DRW interview.

Chapter 22

1. Walter C. Williams, interview by Addison M. Rothrock, Jay Holmes and Eugene M. Emme, John F. Kennedy Oral History Collection, March 25, 1964.
2. J. R. Dempsey, W. A. Davis, A. S. Crossfield, and Walter C. Williams, "Program Management in Design and Development," in Third Annual Aerospace Reliability and Maintainability Conference, Society of Automotive Engineers, 1964.
3. Abbey, interview, 2020.
4. Bruce Tuckman, "Developmental Sequence in Small Groups," Group Facilitation: A Research and Applications Journal, 63(6): 71–72, 1965.

5. Hamish Lindsay, "Apollo 13: 'Houston We've Had a Problem,'" April 15, 2020.

6. David R. Williams and Brian J. Jones, "EMU Shoulder Injury Tiger Team Report," September 2003, NASA/TM-2003-212058.

7. Shane Springer, "Tigers in the Office," Medium.com, July 17, 2019, https://medium.com/swth/tiger-team-for-business-efficiency-30a47a3bf4c6.

Chapter 23

1. Griffin, DRW interview.

2. Griffin, DRW interview.

3. Jefferson Howell, interview by DRW 2020.

4. Griffin, DRW interview.

5. Rothenberg, DRW interview.

6. Kathy Nado, interview by DRW, 2020.

7. Rothenberg, DRW interview.

8. Janet L. Kavandi, interview by DRW, 2020.

9. Edgar Schein, "A Culture Discussion with Edgar Schein," interview by Tim Kuppler, CultureUniversity.com, January 10, 2016, video, https://www.youtube.com/watch?v=gPqz315HSdg.

10. John Maxwell, "The Problem of Pride," January 22, 2014, https://www.johnmaxwell.com/blog/the-problem-of-pride/.

11. Edgar Schein, "Humble Leadership," presentation at Google, February 2, 2016, video, https://www.youtube.com/watch?v=6wJaNKIALLw.

12. Wetherbee, DRW interview.

13. Wetherbee, DRW interview.

INDEX